C

SEE, HEAR, INTERACT:

Beginning Developments in Two-way Television

by
MAXINE HOLMES JONES

The Scarecrow Press, Inc.
Metuchen, N.J., & London
1985

Library of Congress Cataloging in Publication Data

Jones, Maxine Holmes, 1929–
  See, hear, interact.

  Bibliography: p.
  Includes index.
  1. Television in education--United States--Case
studies. 2. Closed circuit television--United States--
Case studies. I. Title.
LB1044.7.J66 1985   371.3'358   84-10715
ISBN 0-8108-1720-9

Dedicated to the memory of my parents,
Clara and Clay Holmes,
who encouraged curiosity and
nurtured my joy in learning

and

to my sister Ginger,
who broadened my horizons by providing
new experiences and fresh points of view
with laughter and with love

CONTENTS

ACKNOWLEDGMENTS

☐   Many people contributed to the development of this book.
I first heard the phrase "two-way tv" through Welda Swed and
Bill Langbehn in the fall of 1977 in the Old Radio Hall of the
University of Wisconsin-Madison. The phrase, and the information related to Trempealeau County, settled into a neuron in
my brain to incubate for a year or so. Then, the concept began
to seem important, and in due course, consumed much of my
intellectual curiosity. Dr. Ann D. Becker first suggested I
keep complete notes, that the information might build into some-
thing worthwhile. Dianne Williams Hopkins, Rhonda Robinson,
Paul Edison-Swift, Marina McIsaac, Michael Streibel, Bob Fina,
Lowell Wilson, and Murray Cornwell shared ideas, knowledge,
reference citations, and encouragement along the way. Dr.
Hal Riehle, Director of the Bureau of Audiovisual Instruction,
and the faculty and staff at BAVI, endured my single-topic
conversations, and somewhat erratic schedule, with tolerance,
patience, and stability. Dr. David Barnard, the faculty at
the University of Wisconsin-Stout, Ralph Whiting and Richard
Sorenson of the Wisconsin Department of Public Instruction,
all generously shared knowledge and helpful background information in many enjoyable interviews and conferences. Nancy
Fowell, Ginny Ward, Irene Stoller, and Deane Hill Jacobs
offered help and cheering comments during many low moments
of frustration. Betty Nixon epitomized friendship in providing a very necessary haven for laughter, joy, and change-
of-pace. To all of them I offer a very humble "thank you."
Words really cannot completely express my appreciation for
all the indications of support from my friends in Wisconsin.

Information for each television project was provided
through directors and coordinators working at each site.
Their willingness to share information and provide additional contacts broadened the content of this text. In all
cases, these busy people were willing to make time to talk,
share, and assist in the clarity and correctness of the manu-
script. To all of them, listed in the many reference

vii

sections and in Appendix C, I sincerely offer my appreciation and gratitude.

In addition, without the gentle assistance of Archie Dixon and his "Place" in Broken Arrow, the stacks of rough draft pages would never have been typed. As always, Jan, Rick, John, Erick, and Cathy offered empathy and reassurance in large, continuing doses. Dianne and Deane deserve a special "thank you" for continuing to inquire and gently nudge this work along. They sensed the support I needed through many bleak and negative moments. Megan Elizabeth, in her own delightful way, increased in wisdom and strength as the pages of this volume began to take shape. Her growth, as well as the arrival of Cassandra Marie and Jacob Allen, recharged my tired efforts toward the completion of the task. In one sense this book is for them, and for all small children in the world today. Electronic communications that seem experimental now will be a commonplace part of the everyday world in the future. Ever the optimist, I hope electronic technology will help all people everywhere, really communicate, understand each other, and move toward peaceful, productive societal interaction.

INTRODUCTION

☐   The people of the United States are in the process of social change caused in part by the lessening supply of oil and gasoline due to forces outside individual or government control. Americans have become an interactive, mobile people and change that inhibits or prevents socialization or new vistas to explore will be met with much resistance. One alternative which could be more completely explored in the coming decades would be the development of interactive television for business, education, industry, and home use. Social contacts could still occur in a slightly removed, but equally interesting and appealing, format. Vistas for exploration via television could be increased and refined in a manner not yet used in the traditional television medium.

This book will seek to explore the potential of interactive television by providing examples of prototype systems which have already been explored. The technology for this medium has already been investigated, at least with experimental equipment, procedures, and audiences. Readers can study, observe, contemplate, and determine if any of these test situations should be examined in greater depth, with more complete analysis and with larger applications for the total population.

As the cost of energy sources for transportation in America becomes more prohibitive, the demand for electronic communication is increasing. Electronic communication exists in many forms--telephone, electric mail, electronic data bases, video (via over-the-air, cable, both open and closed, and satellite transmissions), and radio (AM, FM, amateur, CB, land mobile)--the list is long. This work will deal primarily with communication that provides a visual component and some form of live (real-time) return message from the audience members. Two-way television (interactive television) does not require video at both "ends" of the audience; however, some form of message sending from

ix

the audience must be possible during the actual presentation
or program. Delayed return is really just one-way transmission; the response systems are not able to take place
during the presentation. Questions and answers are therefore reflective, not generated immediately by the dynamic of
interactive communication.

Experimental programs, projects, and test situations
have been conducted by several broad groups of private business, federal government agencies, and research organizations during the past ten to fifteen years. Through these
various test situations, technical transmission, equipment
accuracy, quality of send and return messages, audience
reactions, and social communication have all been studied.
These short-term programs have contributed to the growing
body of information concerned with creating the most efficient and practical method of providing interactive television for the general population.

This research, or information, is also providing, to
some degree, data on the interaction of people and machines,
Primarily, it is the interaction of person with person via a
machine, albeit a highly sophisticated one. The interactive
television systems that use computer response programming
are still person-to-person interaction with a greater use of
sophisticated technology. After all, despite some sense of
wonder (and irritation), a computer cannot program "out"
what some human brain has not programmed "in." Language
and terms of communication may be more tightly controlled,
but the human interaction is still present, coupled with additional equipment.

See, Hear, Interact presents background information
on experimental studies in several fields: education, for
young people and for adults; social services; telemedicine;
citizen participation; in-service information; and community
services. This sampling of prototype interactive systems
provides a view of everyday services which may be available in the future. Technology, consumer interest, government support, industry, and business will all be important
in determining the speed and completeness of development of
these services throughout the nation. Consumer interest and
practical business considerations will in large part determine
the growth, transmission quality, and uses for interactive
television in the coming decades.

SEE, HEAR, INTERACT:

BEGINNING DEVELOPMENTS IN
TWO-WAY TELEVISION

I. MORNING SUN

☐   The public schools in Morning Sun, Waco, Wapello,
and Winifield, located in southeastern Iowa, share learning
through the transmission of microwave television signals.
The idea was conceived in 1978 by Fran Davis, then Director of Education for the Area 16 Intermediate Education
Agency. The idea was discussed with Dale Hemmie, President of Telecom Engineering, and within two weeks Telecom
had designed the system and the equipment for shared visual
projection.

Mr. Davis approached the Iowa Department of Public
Instruction, explained the concept, and through the cooperative
efforts of the school administrators of Morning Sun, Waco,
Wapello, and Winifield, was awarded a Title IV Grant for
$220,000. One hundred thousand dollars of this grant was
funded in the first year in order to develop the equipment
and create an operational transmission system. In the fall
of 1980 each school district entered into a pre-construction
agreement with Telecom Engineering.

Pre-construction activities, after the Title IV Grant
was obtained, included Federal Communications Commission
approval, given within six months. Since microwave transmission requires a direct line-of-sight for optimum performance, topographical maps were used to determine necessary
antenna heights.

Television transmission is totally by closed circuit
microwave signals. Six-foot dishes send and receive video
and audio signals between the four cooperating schools. The
Microwaves for Learning transmission system designed by
Telecom Engineering provides visual and audio reception from
three distant locations. Each school has the capability of
originating a class or receiving a class. When receiving a
class, the students see and hear the instructor on a large
television set or monitor. When originating a class, the instructor sees and hears the three schools on a special television console that has a split screen. This system allows

all classes to be seen without visual façade or "jumping" from one television screen to another, and without exclusion of any location at any time. Audio transmission is boomed into each television classroom so that individual microphones are not required. Inter-school conversation continues in an ordinary manner. Students adapted to the interactive television classes easily and responded positively.

The school districts range in enrollment from 275 (Morning Sun) to 1,001 (Wapello). Because of this limited enrollment many curriculum offerings had not been feasible. Through the Microwaves for Learning project, each high school has designated one room as a television classroom, and from this television classroom high school students in each school district have had the opportunity to study an enlarged curriculum. Four classes in Spanish, and one each in Creative Writing, Anatomy, Physiology, and Advanced Math were offered during the 1980 school year. Approximately 125 students participated in television classes the first year.

In that first year of television interconnection the four school districts developed a common school calendar, and synchronized daily time schedules for classes and school activities. Therefore, interactive classes in one district coincided with classes scheduled in the other districts. Distribution of learning materials was handled initially through an interschool busing arrangement. In 1982, facsimile equipment was installed that provided almost instantaneous transmission of printed materials to all schools.

Evening classes for adults were begun in 1981, offering credit from Southeastern Community College. The University of Iowa planned to offer graduate study through extension courses via the interactive system. The Iowa Department of Public Instruction has observed the development of the microwave system and expanded use of the facilities is under consideration. Long-range plans may link 13 school districts and the University of Iowa through interactive television communication.

Faced with the economic realities of the early 1980's, especially in the field of public education, administrators of the four school districts in the Microwaves for Learning project considered interactive television classes as a better alternative than consolidation of the school districts. From the beginning of the project all personnel, faculty and students included, were actively involved in the development of the Microwaves for Learning system. Faculty participated in summer workshops designed to evaluate, enlarge, and create interactive course materials. Students gained new socialization skills through interactive television communication. Study groups based on the content of the television courses have developed. From this base, a larger social network of students has evolved.

As the Morning Sun, Waco, Wapello, and Winifield Microwaves for Learning program moves into the third year of operation the future seems promising. Development of a second channel for microwave transmission is under consideration. This would allow fourteen classes to be taught per day via the interactive television system, and provide for a much larger curriculum schedule. Teachers remain enthusiastic about teaching via the interactive television system. Genuine rapport and communication between teachers and students, at all school sites, seem to have developed. Students have gained many skills related to television direction and production, especially in visual transmission, and seem fascinated in learning through this form of technology. Expanded communication skills have developed on many levels for teachers and for students.

Extended uses for the interactive television system include joint school board meetings, programs for gifted and talented students, faculty departmental meetings, meetings for administrators, faculty in-service meetings, adult education, and community service. Interactive television has provided a network for increased communication in southeastern Iowa.

REFERENCES

"Microwaves for Learning." Telecom Engineering, 1980. P.O. Box 108, Fort Madison, IA   52627.

Sweet, Michael. "New Look for Education: Television Technology Triumphs," Hawk Eyeland, Sunday, October 5, 1980. Hawk Eye Newspaper, 800 South Main, Burlington, IA   52601.

Two-way Instructional Television (TWIT): 1980 Student Handbook for Instructional Television Classes in Morning Sun, Waco, Wapello, and Winifield School Districts. Morning Sun Public Schools, Morning Sun, IA   52640.

☐   Eagle Bend, Minnesota, population 557, has been the
primary site for an experimental television project under the
supervision of the Federal Communications Commission (FCC).
This experimental project has assessed the facility and value
of Communicasting, which is a low-power, limited-range,
interactive television transmission system.   Administrators
of Eagle Bend Public School District Number 790 became
aware of the 1978 FCC request for experimental programs.
These programs were to be designed to test the value of the
community educational radio fixed service (CERFS) frequency.
Limited, noncommercial television programming connected
with educational, health, public service, and community
affairs was indicated as the focus for these experimental
programs.

Eagle Bend school administrators contacted the FCC,
petitioned for permission to develop the communicasting system, and were finally granted FCC permission to develop
this low power frequency as an experimental television system under FCC guidance, inspection, and approval.

The public school districts of Eagle Bend, Clarissa,
and Bertha-Hewitt, Minnesota, submitted a joint proposal to
the Minnesota State Department of Education in January 1979.
This proposal, Communicasting for Educational Purposes
(CEP), was approved in April 1979.   It was funded under a
Title IV-C developmental grant over a period of three years
and was to be reviewed and renewed annually.

The funding of this project allowed the creation of one
of the smallest working television networks in the United
States.   The network operates under Standby Temporary
Authority from the Federal Communications Commission,
renewable every three to six months.   Three school districts
comprise the network, and each school district is separately
licensed by the FCC.

The television station KG2XCB, Channel 45, is classi-
fied by the FCC as experimental and noncommercial.   Only

a limited number of experimental low-power television (LPTV) licenses had been granted by early 1982. A temporary moratorium has been placed on additional licenses to provide a period in which the FCC can gather data on the experimental low-power television stations and develop appropriate regulations.

One major consideration for the local application of this proposed television system was the possibility of providing television programming at home. An over-the-air transmission system would not require any costs to homeowners other than that of a normal UHF antenna; in other words, no cable lines or connections and no cable subscription cost would be needed. This area of Minnesota was able to receive only one commercial broadcast channel; and no cable services, except in the city of Clarissa, have been available in the area. Also, no educational television was available for school use. So, the proposed communicasting system would provide television services for the general public as well as for the educational institutions.

Construction of the transmission system began during the summer of 1979. Faculty interest and involvement began simultaneously. Approval by the FCC was required with every step in the construction and development of the system. Transmission is by open circuit microwave in the 470-to 930-MHz range. Three antennas were required, located at Eagle Bend, at Clarissa (five miles to the southeast), and at Bertha-Hewitt (twelve miles north of Eagle Bend). A radius of fifteen miles is involved. The last antenna was erected in August 1980. Transmission began in September, with a regular broadcast schedule functioning by October 1980.

FCC approval and formal dedication of the television transmission facilities occurred on January 8, 1981. FCC representatives were present, as were Ed Pillar, President of the Communicasting Association of America; Dr. Lee Cohen, Director of the Center for Advanced Study in Education, The Graduate School and University Center of the City University of New York; and local school representatives and community members.

From the beginning of this project each of the participating school districts has had equal representation and voting power. A Joint Powers Board--composed of two representatives of each school district, the superintendent of each school district, and the CEP Project Director--met

frequently and determined the direction of development. A
Citizens' Advisory Board composed of lay people was also
organized. Input and feedback from these groups have en-
abled all of the communities to stay abreast of developments
in the construction and operation of the transmission systems.

Enrollment in the high schools in the CEP Project is
not large; enrollment over the past five years ranges from
300 to 650 students. Providing worthwhile educational opportunities to the students has a high priority in the communities, and the interactive television system has become one
way to broaden the curriculum in this rural area of the
state.

Each high school in the CEP Project has one studio
room used for the televised classes. If the originating class
is in Eagle Bend, transmission to and from the other schools
include video and audio signals. If the originating class is
in Clarissa or Bertha-Hewitt, equipment capability is limited
to audio return only. Classes are conducted in an interactive "live" format. No one-way, pre-recorded instructional
material is used for the high school classes. Films, including educational visual materials, are shown when cleared
by the producers for projection. However, this material is
treated as in a "traditional" classroom, i.e., with introduc-
tion/preparation before viewing and discussion/follow-up
activities after viewing.

School administrators feel the interactive television
system is a tool to provide education to students and they
encouraged the involvement of faculty members from the beginning of the CEP project. The interactive television system is not designed to replace teachers, but functions as an
alternate method of providing an enlarged curriculum selection for students. Classes during the first year of transmission included German and mass communications. The
school districts in the CEP project had not been able to provide these courses in each of the three schools in previous years.
During the summer of 1981 teachers were involved in in-
service preparation of additional courses for the 1981-82
school year. During the 1982 school year, course offerings were expanded to include employment skills, art, and
advanced mathematics.

The Eagle Bend Communicasting for Educational Purposes Project transmits one-way television to the homes in
the area, via over-the-air broadcast. By the spring of 1981,

$12\frac{1}{2}$ hours of television per day were broadcast. All programming, school courses included, is available for home
viewing. Local programming has included local news, live
coverage of high school graduations, and a series of taped
programming of the local Summer Festival. Through arrangement with the Children's Television Workshop, "Sesame
Street," "Electric Company," and "3-2-1 Contact" were
broadcast the first year of operation. Difficulty in obtaining
videotapes of these programs prevented broadcast during
1982. However, plans are under way to secure the necessary permission and technical capability to rebroadcast these
programs from the Public Broadcast System in Appleton,
Minnesota. Broadcasting arrangements with the independent
network news, religious television channels, Appalachian
Educational Network, and the University of Georgia have
broadened the programming available and allowed initiation
of adult education services via television into the area.
Parenting, pre-school, and pre-GED programs have also been
included in programming schedules.

A survey was sent to viewers in the spring of 1981
and 10 percent of the surveys were returned. Only two surveys were completely negative; the rest of the surveys contained some measure of positive response. The most fre-
quent comment was a request for more detailed time schedules for programming, and longer advance notice for scheduled programs. In short, it seemed the citizens of central
Minnesota were interested in the low-power television system and wanted to use the information and recreation available to them through this system.

By the summer of 1981, 50 high school students were
trained and were able to assist in the operation of the television station. Many skills in a broad variety of tasks are
required within a television station. Not all the students
mastered all the specific skills needed in a television station within the first eight months of broadcast operation.
However, the opportunities for student participation in many
phases of the programming and operation of this low-power
communicasting have provided an entirely new "hands-on"
experience for them. Career possibilities that students had
generally not considered in the past are now available for
realistic appraisal through this high school study/experience.

In December 1981 a preliminary report, the first in a
series of three, on the Communicasting for Educational Purposes Project was released. This report was designed to

convey to funding sources a complete description of project implementation to that time and a review of the critical issues involved in operation of the program.

The report indicated several areas of concern. There
appeared to be confused communication regarding decision-
making and equal involvement in the television project. Administrators of Clarissa and Bertha-Hewitt felt left out of dis-
cussions and decisions in implementing and operating the television system. Joint Powers Board members perceived them-
selves as advisors not involved with actual decisions. Faculty
members desired more complete in-service training for continuing television teachers as well as for those teaching by television for the first time. The faculty also wanted to receive
tangible rewards for teaching via the two-way television system.
Some concern existed regarding the open-circuit television sys-
tem--the faculty were hesitant to teach "in full view" of the
community. Also, additional assistance was desired in all
schools in encouraging students to enroll in the two-way television classes. The faculty indicated that students in the remote classes should have adult supervision and that administrators should be aware that the tendency to "hand-pick" students carefully for the television courses might prevent some
students from obtaining needed courses. Lastly, the cable firm
operating in Clarissa had failed to provide subscribers with the
necessary UHF equipment to receive Channel 45. This action
prevented a sector of the project community from participating
in the CEP programming at all.

However, this same report highly praised the Communicasting for Educational Purpose Project. The fact that the system functions at all amazes some experts. The variety and
length of daily service is commendable and the implementation
of two-way classes is an accomplished fact. Only two service
interruptions have been recorded; one was caused by lightning
and the other by mechanical functions.

The costs thus far seem high to those outside the field.
Knowledgeable experts feel the system is functioning at a
fourth of the cost of a UHF broadcasting station, which the
communicasting staff had not intended to replicate.

Additional developments during 1981-82 included adding
a teacher representative from each school district to the Joint
Powers Board, and expansion of air time to an average of 14
hours per day. Data indicated the following: the station could
operate efficiently with two full-time positions, plus one half-

day position during the normal eight-hour working day; one hour
of programming required approximately one hour and forty
minutes of production time; teachers can teach effectively in the
television system; students can learn effectively; and, the equipment can function as it was designed to function.

In 1982 CEP administrators were looking forward to
offering a full-day schedule of classes via television and adding
three additional school districts to the television network.
Cable companies were to be requested to add CEP programming
to their system so that the Communicasting for Educational
Purposes Project could reach a larger number of communities
in rural Minnesota.

The school districts of Eagle Bend, Clarissa, and Bertha-Hewitt, through the joint CEP project, have been charting
unknown waters in the field of low-power television. This transmission system has been relatively unknown in the industry, as
the experimental site stipulations by the FCC have indicated.
Experimental programs test new equipment and new ideas;
frequently, rough sailing is involved in the process of development and implementation. The value of experimental programs
is found in the long-range benefits, determined after a period
of evaluation, revision, and adjustment. The results of the CEP
Project, initiated by individuals in the local school districts,
will assist in indicating the feasibility of using this low-power,
limited-range television transmission system for educational,
institutional, and local community purposes during the coming
decades. Eagle Bend, Clarissa, and Bertha-Hewitt school
districts are providing the information needed for experiments
to become everyday occurrences.

REFERENCES

Cohen, Lee, and A. Edwin Piller. Petition for Rulemaking:
     In the Matter of the Establishment of a New Community
     Educational Fixed Radio Service in the 470MHz to 930MHz
     Frequency Band, January 21, 1977. The Graduate
     School and University Center of the City University of
     New York, Center for Advanced Study in Education,
     Institute for Research and Development in Occupational
     Education, 33 West 42 Street, New York, NY 10036.

Communicasting for Educational Purposes. Title IV-C
     Developmental Grant submitted to the Minnesota State
     Department of Education, January 31, 1979. This was

a joint proposal submitted by Eagle Bend School District # 790; Clarissa Public Schools, Independent School District # 789; and Bertha-Hewitt Independent School District # 786. Contact Eagle Bend School District # 790, Eagle Bend, MN 56446.

"FCC Urged to Allocate 'Communicasting' Service Frequencies," Microwaves, March 1977. Hayden Publishing Company, 50 Essex Street, Rochelle Park, NJ 07662.

Jones, Maxine. "Communicasting: Using Television for Community Services," Wisconsin Library Bulletin, 76(6), (November/December 1980): 270-72. Department of Public Instruction, State of Wisconsin, 125 South Webster Street, P. O. Box 7841, Madison, WI 53707.

Morehouse, Diane L. Communicasting for Educational Purposes. A Project Funded by: ESEA IV-C and the Council on Quality Education; A Preliminary Evaluation. Evaluation Section, Division of Special Services, State of Minnesota, Department of Education, December 31, 1981. Educational Media Unit, Division of Instruction, Minnesota State Department of Education, Capitol Square Building, St. Paul, MN 55101.

"New Community Educational Fixed Radio Service and Limited Non-commercial Local Origin Through Television Translator Stations; Federal Communications Commission; Released August 22, 1978," Federal Register, 43(166), (August 25, 1978):3808+. Supt. of Documents, U. S. Printing Office, Washington, DC 20402.

"Petition Filed with FCC to Allocate Education Band," Electronic Engineering Times, February 21, 1977. CMP Publications Inc., 111 East Shore Road, Manhasset, NY 11030.

Piller, S. Edwin. Communicasting, A New Low Cost Community Telecommunications System for Science, Education, and Public Service. Paper presented at the National Telecommunications Conference, June 18, 1979. Contact: S. Edwin Piller, President, Communicasting Association of America, Inc., Syosset, NY.

III. TREMPEALEAU COUNTY

☐   Trempealeau County, Wisconsin, is a sparsely settled rural area. Reports in 1970 indicated that the total population was less than 24,000, 23 percent of the population was under twenty years of age, and 16 percent of the population was over sixty-five. Median income in 1970 was about $7,400. At the same time, 14 percent of the families had incomes below the poverty level. Twenty-six percent of the work force was involved with agriculture. Yet, despite this rural setting, Trempealeau County contains one of the most progressive television systems in the country: a county-wide, two-way television response system designed to provide, eventually, all subscribers with a variety of electronic services.

What triggered the development of this sophisticated system in peaceful, placid, western Wisconsin? One must look into past events in rural Wisconsin and learn that Trempealeau County has a half-century history of cooperative efforts. Developing cable television within the area was discussed over fifteen years ago, and a logical method was to consider a cooperative development, actively involving rural residents, similar to the procedure used in installing electricity throughout the county during the 1930's. Data indicated that a large percentage of the population was home during the day; television sets were in 93 percent of the homes and there was high television usage, despite reception from only two television channels. Civic leaders living in the Trempealeau County area sensed the potential usefulness of cable television for county residents, often isolated by geography and climate.

In 1973, citizens banded together with the encouragement of Gordon Meistad, Manager of the Trempealeau Electric Cooperative, and purchased "preferred equity certificates" to indicate citizen support for the two-way television system. From this cooperative effort the Western Wisconsin Communications Cooperative (WWCC) was established. It is a consortium of twenty-three Trempealeau County cooperatives, seven school districts located in Trempealeau County, and one school

district located in Jackson County. WWCC proposed to provide
a broadband communications network accessible to all 9,500
households in the county.

Funding was a primary consideration in the WWCC proposal. The engineering consulting firm of Ralph Evans and
Associates estimated the cost of initial implementation of the
system at $1,245,000. Application was made for a Com-
munity Facility Loan from the Farmers Home Administration
under the Rural Development Act of 1972. The Farmers
Home Administration identified twenty conditions that WWCC
would need to meet to qualify for the loan. When these conditions were agreed upon, a loan for $1,440,000 was arrang-
ed. This act in itself makes the Trempealeau County project a unique situation in that this was the only Community
Facility Loan granted under Title I of the Rural Development
Act of 1972 for the purpose of establishing a broadband
communication system. The Farmers Home Administration
required 40 percent of the residents in the county to provide
financial support, many years before receiving services, in
order to secure final approval by government agencies (hence
the value of the "preferred equity certificates"). Face-to-
face interaction, explanation, and discussion with many
county residents was required in order to gain the understanding and support needed for the proposed cable system.
Construction delays caused by the precise requirements and/
or additional stipulations of the Farmers Home Administration, the exact specifications of construction, the custom-
ary delays that occur in building procedures, weather and
geographical constraints, and the huge land distance to be
covered for this project (including the installation of 129
miles of cable transmission lines) all contributed to a later
operational date than originally anticipated. Many installation difficulties and delays occurred. Throughout this
project, the cooperation, understanding, and support of the
county-wide community was desired and developed. In subsequent conferences, seventeen additional conditions were
required by the FHA. WWCC met these conditions within
the allotted time span. Much effort and energy was expanded
in complying with the tremendous amount of detail and uni-
formity required by the Federal Agency. The Farmers Home
Administration Loan was finalized on November 15, 1978.

Another cooperative group was formed composed of
eight public school districts and four member universities
of the University of Wisconsin system. This organization,
the West Central Wisconsin Consortium (WCWC) investigated

ways to develop the two-way cable system for cooperative use
by the eight school districts. Application for funding was
made to the W. K. Kellogg Foundation in Battlecreek, Michigan. In October 1976 a four-year grant of $510,170 was
awarded. This educational component of the county two-way
cable television system then became known as the Trempealeau County Kellogg Project.

The Trempealeau County two-way cable project has
been unusual for many reasons. Construction and implementation of equipment has required cooperation among a
wide variety of individuals, businesses, and groups. Two
independent cable companies were operating in Trempealeau
County when the two-way project began, so it was necessary
to blend the franchise area of each operator to include all
of the school districts.   And, as in most states, the Wisconsin school districts were restricted by law to financial
agreements of three years or less.   State legislation was
changed to allow the schools to participate in a fifteen-year
lease arrangement with the cable companies, and state and
national legislation needed to be passed to allow for the long-
term financial agreements between the school districts and
the Farmers Home Administration.   Legislation was also
required to allow the Western Wisconsin Communications
Cooperative to collect the necessary local funding.   As a
result of the Trempealeau County efforts, state legislation
was developed to allow intercommunity cable districts to
organize and use municipal bonds for financial banking.   Discussion and agreement was needed on local, state, and national levels.

The Trempealeau County two-way television project
proceeded in three primary phases of development. Phase I
included cable interconnection for the schools and homes in
the larger communities of the county.   Approximately 2,900
homes and 230 businesses and institutions were connected
during Phase I.   The eight high school buildings were interconnected by two-way video and audio transmission.

Phase II expanded the cable system to the less densely
populated areas of the county and to the smaller villages. A
$4.13 million loan from the Rural Electrification Administration
completed Phase II.   This allowed cable services to reach
52 percent of the homes in Trempealeau County.   Eleven
cable franchises now operating in the county will be aug-
mented by two more franchises.   Over 500 miles of cable
will have been installed in the county, with 75 percent of

the cable placed under ground.   As part of the community
services provided to this rural population, a television studio, funded by a National Telecommunications Information
Administration Grant of $45,000 with local matching funds of
$15,000 has been located in the county courthouse in White-
hall.   This studio functions as a local access center for
information programs on health, social security, concerns
of the aging, local news, regional interests, and county
government meetings.   Videotapes have been produced locally,
concerned with issues of health, aging, and local interests.

        Phase III will connect the most isolated areas of the
county, and the remaining 48 percent of the residences.
Connections will include fire alarms (particularly useful for
barns) and medical alert signals.   There has been discussion
about other consumer services, including banking and shopping by television.   The increased quantity of network television channels, independent cable channels, channels for
local community interests, school classes, and general information now available has greatly broadened the viewing awareness and opportunities for cable subscribers throughout
Trempealeau County.

Trempealeau County Kellogg Project

        Schools in Trempealeau County have small enroll-
ments, ranging from 390 students (K-12) in the smallest
school to just under 1,600 (K-12) in the largest school.
Fewer than 7,000 students are enrolled in all of the eight
schools in the two-way television system.   There are approximately 2,500 high school students eligible for curriculum
over the two-way television system.   A broad-based curriculum for each school district has been almost impossible in
the past because of financial constraints.   However, the
school building is still considered the hub of activities for
the small communities of the county, and these communities
want to retain their local identity.

        In 1972, five of the school districts developed a co-
operative purchasing procedure which was broadened a few
years later to include cooperative education, accomplished
by busing students to the neighboring school where a special
course, such as French, was taught.   Because some of the
school districts are over sixty miles apart, not all schools
in the county were able to participate in this form of cooperative education.

The small budgets, large distances, and limited curriculums of the past provided a situation in which two-way interactive cable television could provide a viable alternative through which students could gain the educational skills in the 1980's and beyond. Two-way television allows students to be connected visually and audibly, without the time, effort, and expense of busing. In addition, adult community interest programs became available for everyone in the county almost immediately. However, the medium, methodology, and equipment were untried, unknown, and frightening to some individuals when the educational aspect of cable television was first discussed in the early 1970's.

Mention of two-way television to teachers in Trempealeau County during the early stages of the cable television development caused many mixed reactions. Teachers felt threatened and frightened; they also felt that the television system theoretically should increase the status and opportunities of faculty members. A variety of very human emotions have been aroused in members of every school district whenever major audiovisual changes have been implemented. Change of any sort often causes fear of the unknown--fear of a situation not previously experienced.

To address these concerns of faculty members and to explore the educational opportunities, the West Central Wisconsin Consortium (WCWC) was formed prior to 1975. Members included the University of Wisconsin--Eau Claire; the University of Wisconsin--LaCrosse; the University of Wisconsin--River Falls; and the University of Wisconsin--Stout. WCWC developed "Restructuring Rural Education: New Delivery Systems and Curriculum Formats" during 1975. The proposal listed eleven broad educational objectives ranging from pre-school through adult areas of interest.

In 1976 a grant from the W. K. Kellogg Foundation, Battlecreek, Michigan, provided $578,170 to be allotted over a four-year period. This money allowed each of the school districts to acquire basic television equipment. This "start-up pack" included three one-half-inch Sony videotape recorders and one color camera. Editing equipment was purchased and time-shared between schools. Staff at the University of Wisconsin--Stout designed a custom-built signal frequency channel scanner to allow the eight school districts to signal each other in the event their television monitor was not on or was tuned to another channel. Individual school districts purchased additional television equipment. Later,

in 1980, Eleva-Strum purchased microcomputers, which
were placed in the audiovisual room. These microcomputers
are used for interactive television computer courses and
often provide visual displays and messages to the other
schools. Technical capability in the Trempealeau County
system is designed so that each classroom in the high school
buildings is able to originate two-way classes for television
broadcast.

Requirements of the Kellogg grant included the pro-
vision of adult education for the county. This programming
developed quickly with a leasing arrangement from the court-
house television studio for continuing education and degree-
requirement courses. Area universities, Appalachian Community Service Network programs, and local programming
provide credit and noncredit educational programs for the
county.

Distances between school building sites seemed to
make this educational project move slowly. For the initial
phase of development, 129 miles of transmission cable needed to be installed.

In anticipation of completed construction and two-way
television transmission activities, the schools conducted a
workshop in the spring of 1977. This workshop was designed
to develop curriculum that could be used via the interactive
television system. All faculty members of all the school
districts were included. Rapport among department members
of different school districts developed; prior to the workshop,
teachers from different districts were largely unknown to each
other. "Departments" of single teachers cannot really develop an exchange of ideas. However, combining the faculty
of all the districts did provide a group of teachers for each
department, and ideas and mutual efforts toward effective
two-way television teaching developed. A core, basic curriculum to be used by all eight participating school districts
emerged from the communication started during this workshop.

During the period 1978-79, semimonthly meetings were
held for the "television representatives" from each school
district. Information gained at these meetings was dissemi-
nated to the individual school faculties. Throughout the curriculum development and in-service meetings leadership
philosophy emphasized the following: 1) two-way television
would not replace the teacher; 2) two-way television is to
assist the teacher on a day-to-day basis; 3) planning should

be based on the year-long curriculum; and 4) the faculty
should determine how the two-way television facilities can
aid the teachers in a continuing manner.

Curriculum activities began with a series of mini-
courses designed to acquaint students and faculty with the
unique characteristics of interactive video/audio television
communication. Course offerings were broadened the second
semester to include music theory, advanced mathematics,
computer science, and Spanish. Programs from the Appalachian Community Services Network were examined, and some
local production of educational materials developed during the
first year of transmission.

The Trempealeau County school cable television signals are transmitted primarily by cable, with a microwave
transmission link between the north, central, and southern
portions of the county. Cable services to the schools include all of the channel offerings available to the homes except entertainment movies. Transmission signals between
the eight high schools are two-way video and audio. The
school channel at present is open-circuit; and the equipment is
designed to allow for switching to closed-circuit transmission
at some future time. The school channels are in the eight
midband channels that can connect all of the school television
production areas simultaneously. The high school buildings
are wired to use any classroom as a two-way television
classroom production area, with transmission to other high
schools classrooms in the Trempealeau County project. Faculty members consider classrooms as "two-way areas," not
as television "studios." Television transmission provides an
extension of the preparation, planning, and implementation
needed when teaching students within a single classroom.
Faculty are given financial remuneration if the two-way class
preparation/production is above the normal work load.

Installation of transmission facilities for the school
interconnect was handled by a turn-key contract. Initial
transmission occurred in July 1979 and in December 1979,
the first complete two-way interconnect occurred. On
March 13, 1980, the installation of facilities for the two-way
interconnection of all high schools was completed. Interconnection activities for conferences for administrators, interschool departmental meetings, citizens groups, county
agents, and social services occurred during the following
year.

.Each school district has an ongoing fifteen-year lease
with the cable cooperative. For the first three years of
development of the two-way television system the schools had
a reduced payment of $4,500 per year. For the remaining
twelve years of the contract the lease rate will be $9,000
per year. Each school district also provided a $1,000 mem-
bership fee in the early stages of organizing the two-way
television project.

Trempealeau County Project CIRCUIT

WCWC submitted a proposal, Project CIRCUIT, to
the Wisconsin Department of Public Instruction in 1980. This
proposal was awarded and funded through a three-year Title
IV-C developmental grant in the amount of $206,000. The
proposal title, Project CIRCUIT, is an acronym for "Curriculum Improvement Resulting from the Creative Utilization
of Instructional Two-way Television." Major purposes of
Project CIRCUIT were to integrate use of the interactive
television system through development and adaptation of curriculum materials to the unique interactive television setting
and to assist the faculty in this adaptive process.

A project director was selected. Philosophies of each
of the school districts were appraised. Evaluations, needs
assessments, and interviews were conducted in each of the
school districts with the data gathered from principals, guidance counselors, and teachers. Written needs assessments
were compiled for high schools; a more inclusive survey of
needs for kindergarten through twelfth grade was collected.
Analysis of these data resulted in focusing on the development and adaptation of the high school curriculum for the
interactive television setting.

Individual faculty members were involved in presenting pilot courses over the interactive television system. The
programming of each subject was considered an experimental
situation, i.e., a teaching method to be tested, evaluated,
and revised as needed. From the first course, the teaching
of a foreign language, developed one faculty communication
preference. The teacher wanted eye contact with all groups
of students at the same time. Rather than switching the converter from site to site, the teacher preferred a television
monitor transmitting visuals from each class site, functioning at all times. In other words, the traditional eye contact
procedures, similar to having the students all located in the

same room, provided a familiar and traditional situation for teaching a foreign language.

During the first year of transmission, team teaching in history, shared between two faculty members, was explored. This course proved very successful since each teacher concentrated on areas of personal preference, and a wealth of learning materials were available to students. The experience eventually indicated that teams of two could successfully teach via interactive television. However, another subject, with instructional duties divided among ten faculty members, resulted in difficulty. A lack of communication concerning specific course materials occurred. Review and inclusion of additional information for continuity of the subject matter presented to students became necessary. Throughout the year guest speakers, clinicians, and field trips were shared via interactive television, with every student able to see and hear from a "first row seat. "

Administrators of Project CIRCUIT are willing to try almost any educational idea a teacher wishes to pursue. However, there is a desire to offer courses which will build interest in further learning. Sequential courses will be provided to allow the interest and motivation gained in the first course to be pursued through additional study.

In 1982 eight credit-courses were offered in mathematics IV, computer science, shorthand, and Spanish I. In the following year, second-year courses for the subjects were offered. Additional subjects that may be offered include music theory, art history, art appreciation, French, German, and woodwind, brass, and instrumental music.

From the early course offerings a nucleus of expertise developed. Early teachers developed experience in course preparation and execution. Subsequent teachers gained from this experience in many ways, including proper camera placement, verbal interaction and involvement, and adaptation of teaching styles from the traditional classroom. "The two-way television system is the next best thing to being there" has been stated by several faculty members. The opportunity to see students via the television monitor provides a greater feeling of involvement, learning, and interaction. Using a telephone network--with only audio contact--does not create the feeling of learning and interaction desired in Trempealeau County. Faculty members are discovering that

anything used in a traditional classroom may be used for
teaching over two-way television.

Experts agree that interactive television should not be
used to substitute for a teacher.   Use of the system allows
for curriculum offerings previously not available within these
small school districts.   Learning experiences formerly impossible because of geographical distances and time constraints
can be provided to students, greatly enriching the educational
information and broadening the intellectual horizons of the
student body.   Interactive television can, in many ways,
eliminate constraints of time and location to provide an almost "on-the-spot" learning experience.

Experience within the Trempealeau County two-way
television project has indicated that the process of adjusting
to teaching by interactive television is really very short.
Teachers involved in the early interactive television classes
had taught from as many as 17 years to as few as two or
three years.   The graphics, visuals, and film used in traditional classes may be used equally well in two-way settings.
Prior planning allows materials to be properly transmitted
via the television camera.   The shorthand materials tra-
ditionally written on a chalkboard were written on a steno
pad, and transmitted to interactive classes via a camera
focused over the shoulder of the teacher.

Early evaluation reports indicate that students learn
as well in television classes as in traditional class settings.
This seems to be true whether students are in the on-site
television class or in the remote-site classroom.   In fact,
there is some indication that remote-site students have higher grade results.   Evaluation procedures now include standardized pre- and post-testing for several courses, which
will provide hard data for long-range evaluation.   However,
the continuing motivation to participate in the interactive
television classes and activities indicates that students and
teachers are gaining knowledge and enjoying the interactive
communication experience.

One use of the interactive television system has been
the development of a Wisconsin area High Quiz, similar to
the "College Bowl" programs over broadcast television many
years ago.   The High Quiz activity has encouraged interaction among schools, is followed by the community, and has
found interest among the junior high students.   Subject areas
include general information, sports, sciences, mathematics,
literature, history, and current events.   Different quiz

sessions, with junior and senior high students, are televised six to eight times a month.

Future plans for Project CIRCUIT include the acqui-
sition of data transmission equipment. Television technology
is available to tie the schools together almost instantly
through the transfer of necessary written materials. Effective feedback of tests and course materials will be expedited
through this process.

Trempealeau County administrators realize the two-
way television technology available in the Project CIRCUIT
schools is far from ordinary. Developments in Trempealeau
County over the past decade may prove useful for other educational and business institutions. Those closest to the
project feel interactive television technology can adapt to the
imagination of the user. The Trempealeau County cable
television project is an example of sustained interest and
effort in sharing a dream. Local individuals banded together
in actively pursuing a dream of instant communication for
every member of the county-wide community. And the dream
is coming true, in rural, western Wisconsin.

REFERENCES

The Feasibility and Value of Broadband Communications in
     Rural Areas. A Preliminary Evaluation, April 1976.
     United States Congress Office of Technology Assessment,
     600 Pennsylvania Avenue, S. E. Washington, DC 20510.

Goldmark, Peter. "A New Rural Society on Telecommuni-
     cations," Rural Electrification, May 1974. (Rural Elec-
     trification is now entitled R E Magazine published by
     the National Rural Electric Cooperative Association,
     Massachusetts Avenue, N. W. , Washington, DC 20036.)

Hoy, Tom. "Rural Communications Co-op: Of the People,
     By the People, For the People," Rural Electrification,
     May 1974. (Rural Electrification is now entitled R E
     Magazine published by the National Rural Electric Coop-
     erative Association, Massachusetts Avenue, N. W. , Wash-
     ington, DC 20036.)

"Restructuring Rural Education: New Delivery Systems and
     Curriculum Formats," West Central Wisconsin Consor-
     tium, 1975. Trempealeau County Cable Project, WCWC,

205 Osseo Road, P. O. Box 326, Independence, WI
54747.

Rivkin, Steven R.   "Bringing Broadband Communications to
    the Countryside," Rural Electrification, May 1974.
    (Rural Electrification is now entitled R E Magazine pub-
    lished by the National Rural Electric Cooperative Associ-
    ation, Massachusetts Avenue, N.W., Washington, DC
    20036.)

IV. INTERACTIVE CLASSROOM TELEVISION SYSTEM

☐   An innovative, closed circuit television system for
handicapped individuals has been developed by the Rand Corporation. Funded in part by the Social and Rehabilitation
Services of the United States Department of Health, Education, and Welfare, prototype equipment was tested and
improved. Commercial production of Interactive Classroom
Television Systems (ICTS) has developed, and this unique aid
for visually and hearing impaired individuals is now available
in libraries and other institutions as well as in home, school,
and employment settings.

The ICTS equipment magnifies written or pictorial
material, which is then viewed via a television monitor by
the visually impaired individual. The magnified picture can
also be projected to a large television monitor for class or
group sharing.

An Interactive Classroom Television System (ICTS)
was tested in a classroom situation with students in Madison
School (Santa Monica Unified School District, Santa Monica,
California), during the 1973-74 school year. A primary aid
for learning was noted in that partially sighted students could,
via the room television camera and screen, see what was
discussed as the material was placed on the class chalkboard, not after the class was completed by going to the
chalkboard to examine it visually at close range. Exchange
of information between students and teachers occurred more
rapidly, and the students felt involved during the entire class
session because they were more actively using their sense
of sight.

A second pilot program was conducted in the Rowland
Unified School District at Killian Elementary School (Rowland
Heights, California) beginning in November 1975. There, an
improved model of the equipment used in Santa Monica was
tested. An enlarged push-button response was developed; the
master control unit was given more flexibility; split-screen
and full-screen superposition capabilities were again used,

and a maintenance manual for use by the faculty was developed. Other improvements included the addition of an audio track fed into earphones and adjusted as needed for each participant. Student responses were similar to the responses in the first pilot program. The additional controls, providing privacy in audio connections and more button response modes, seemed helpful.

One component of the research conducted with the Killian Elementary School students involved the detection and recognition of facial expression. Partially-sighted individuals have difficulty observing facial gestures (raised eyebrows, frowns, smiles, etc.) which normally-sighted persons observe and incorporate into communication behaviors. Consequently, facial expressions are usually not included in the communication and social skills of partially-sighted people. The Rand Corporation research included teaching the seven basic facial expressions: anger, surprise, fear, disgust, happiness, sadness, and neutral; these expressions are all internationally recognized and accepted. To aid the students, these facial expressions were televised and enlarged; students could see the expressions and incorporate meaning for the expression into specific social situations. Students practiced using these facial expressions in social interactions, and seemed to benefit from the information this activity provided. Data from this area of the research project may hold information valuable for sociologists in future studies.

The ICTS equipment for individual use (one station) has several components. A television screen (monitor) for individual viewing is placed next to a television camera. This camera points downward and faces a platform upon which material to be viewed is placed. Materials placed on the platform are televised, via the camera, and displayed on the television monitor. Image enlargement can be adjusted, via a zoom lens, to meet individual needs. ICTS equipment for a classroom situation (in other words, a group of individual stations) requires this same equipment at each station plus a master control unit for the instructor. A room-viewing camera, and a control unit, videotape recorder, and color television monitor/receiver would also be needed. Classroom units also include headphones for private audio connections or for audio amplification.

Difficulties in implementing the ICTS within public schools seem to occur in two main areas. First, school funding is planned on a year-to-year basis. Initial outlay for an ICTS system with eight stations and master teaching unit, though less than $50,000, seems expensive for one twelve-month period. However, pro-rating the cost of this system over the number of years the equipment will be used, including eventual repairs, quickly indicates that ICTS equipment can be both efficient and economical.

Second, faculty training in utilization of the equipment should be an ongoing process, rather than one "hands-on, push-the-buttons" workshop. One strong value of the ICTS system is the visual expansion it provides for students. The ramification of teaching students who can suddenly see more clearly and more completely, cannot be explained in one brief introductory workshop. Ongoing opportunities to examine new teaching materials, new visual experiences, and new teaching methodology, would assist in the continuing utilization and expanding value of the ICTS equipment. The potential value of an ICTS system for the partially sighted seems almost unbounded. Ongoing procedures to explore this potential through workshops and other sharing activities should be made easily available to ICTS faculty.

Response to individual units of ICTS, namely closed-circuit television systems (CCTV) for the partially sighted, has been good. Home use and institutional use in libraries and research centers have provided a market for the single-station CCTV systems. Individual CCTV systems range in cost from $1,400 to $2,800. Specific firms producing the equipment are listed in the Appendix.

One additional note: The staff of St. Mary's School for the Visually Impaired in Dublin, Ireland, built an ICTV system following the directions included in the Rand Publications written by Dr. Samuel Genensky. Difficulties in synchronizing time schedules for the use of the equipment eventually necessitated temporary dismantling of the ICTV station so the individual television monitors could be used elsewhere. However, administrators praised the function and purpose of the ICTV and are in the process of developing ICTV stations for each classroom.

REFERENCES

Bikson, T. K.; T. H. Bikson; and S. M. Genensky. The Impact of Interactive Classroom Television Systems on the Educational Experiences of Severely Visually Impaired

Students. Prepared for the Office of Education, U. S.
Department of Health, Education, and Welfare. Rand
Report R-2395-HEW, Rand Corporation, September 1979.
The Rand Corporation, 1700 Main Street, Santa Monica,
CA   90404.

Genensky, S. M.; P. Baran; H. L. Moshin; and H. Steingold.
A Closed Circuit TV System for the Visually Handicapped.
Rand Memorandum RM-5672-RC, Rand Corporation, Au-
gust 1968. The Rand Corporation, 1700 Main Street,
Santa Monica, CA   90404.

Genensky, S. M.; H. L. Moshin; and H. Steingold. A
Closed Circuit TV System for the Visually Handicapped
and Prospects for Future Research. Rand Paper P-4147,
Rand Corporation, July 1969. The Rand Corporation,
1700 Main Street, Santa Monica, CA   90404.

Genensky, S. M.; H. E. Peterson; R. W. Clewett; and
R. I. Yoshimura. A Second Generation Interactive Classroom Television System for the Partially Sighted. Prepared for the Department of Health, Education, and Welfare. Rand Report R-2138-HEW, Rand Corporation,
June 1977. The Rand Corporation, 1700 Main Street,
Santa Monica, CA   90404.

Genensky, S. M.; H. E. Peterson; R. I. Yoshimura; J. B.
Von Der Lieth; R. W. Clewett; and H. L. Moshin.
Interactive Classroom TV System for the Handicapped.
Prepared for the Department of Health, Education, and
Welfare. Rand Corporation Report R-1537-HEW, Rand
Corporation, June 1974. The Rand Corporation, 1700
Main Street, Santa Monica, CA   90404.

"TV a Boon to Those Who See Poorly," Video Users Market-
place, 11(23), (October 1, 1979):4, 7. Knowledge Industry
Publications, Inc., 701 Westchester Avenue, White
Plains, NY   10604.

"Visual Equality: Interactive Classroom Television," Pre-
sented by the Rand Corporation. Producers, John
Huneck and Thomas Bikson; Project Director, Dr. Sam-
uel Genensky. John Huneck Productions, Crest National
Videotape and Film Laboratories, December 1978.
(Available in 16mm and 3/4 videocassette.)

V. DIGITAL, AUDIO, AND VIDEO INSTRUCTION DEVICE

☐　A Digital, Audio, and Video Instruction Device with
an acronym DAVID has been developed through the cooperation of the National Technical Institute for the Deaf, Rochester Institute of Technology. DAVID has extended the visual screen capabilities of computer-assisted instruction to
include videotape segments using a sense of live interaction.
A variety of conversational responses are stored in the computer. An instructional program begins and the student responds with complete words, phrases, sentences, or other
types of answers, using the keyboard response system. The
computer searches for the appropriate answer, with branching capability based upon the student's input. The audiotape
mechanized conversation is played, the student responds,
and DAVID continues the conversation as the student input
directs. Early software content included simulated job interviews and correct lipreading skills.

DAVID originally focused on the special needs of the
hearing-impaired students. Sample populations of deaf and
normal-hearing students were used in evaluative studies.
The teaching method used interactive television and television.
The media were not compared, but the test results of students were compared. Cognitive recall on memory of diacritical markings was the subject matter. In the first study,
in television format, 10 percent of the students reached
mastery in three hours of time, with very poor retention
over time. In the interactive television format, 87 percent
of the students reached mastery in one and a half hours,
with retention continuous over time. This finding was con-
sistent over the three studies, and no significant differences
in learning were found between deaf and normal-hearing
students.

Two additional pieces of equipment supplement the
original DAVID model. Trainer II is capable of assisting
learning in exactly the same manner as DAVID, minus the
video component. In other words, keyboard responses to
audio signals and computer searches for the proper response

continues.   Audio provides the channel of interactive communication.

VLDS is the language processor used with DAVID. Through the VLDS computer system, programming information is entered automatically without technical computer programming skills.

DAVID technology is available in self-contained units, using videotape or videodisc software.   Instructional materials may also be received through coaxial cable or microwave.

In July 1981, the Veterans Administration began a long-range project to utilize DAVID in applied training for veterans.   Plans include using DAVID to automate the diagnostic procedures of the deaf and to develop complete self-instruction programs for the hearing-impaired.

In the United States, over twenty DAVID units have been placed in industry.   VONTECH, Inc., the production firm for DAVID, has training centers located around the country, providing public service training with job skills, as well as providing training assistance for industry and schools. In addition, Nigeria received DAVID units in 1982, opening avenues of learning for hearing-impaired students and providing assistance in many forms of industrial training.

REFERENCES

Cronin, Barry.   "The DAVID System: The Development of an Interactive Video System at the National Institute for the Deaf," American Annals of the Deaf, 124(5), (September 1979):616-623.   Conference of Educational Administrators Serving the Deaf, 814 Thayer Avenue, Silver Springs, MD   20910.

Smith, Jean McKernan, and James R. von Feldt.   A Comparison of Two Media: An Examination of Computer Assisted and Videotaped Instruction in Teaching Diacritical Markings to Post Secondary Deaf Students, August 1, 1977.   National Technical Institute for the Deaf, Rochester Institute of Technology, Rochester, NY   14623.

von Feldt, James R.   A Description of the DAVID Interactive Instructional Television System and Its Application to

Post High School Education of the Deaf, December 1978
Instructional Television Department, National Technical
Institute for the Deaf, Rochester Institute of Technology,
Rochester, NY   14623.

von Feldt, James R.   An Introduction to Computer Applications in Support of Education, August 1977.   National Technical Institute for the Deaf, Rochester Institute of Technology, Rochester, NY   14623.

von Feldt, James R.   A National Survey of the Use of Computer Assisted Instruction in Schools for the Deaf, January 1978.   National Technical Institute for the Deaf, Rochester Institute of Technology, Rochester, NY   14623.

VI.   ROCKFORD

☐   Rockford, Illinois, became the location for one of the
National Science Foundation projects developed during the
1970's.   Michigan State University, in cooperation with the
Rockford Fire Department and Rockford city administrators,
developed a teaching unit conducted via two-way television.
One of the major purposes of the project was to test the
feasibility of this teaching format and its potential application
to additional instructional materials.

Michigan State University analyzed the potential applications of interactive television for all types of services to
businesses and homes.   Examination of the state-of-the-art
equipment in 1974, when the project began, revealed two
types of cable television.   One was per-channel pay TV; in
other words, a flat rate for so many viewing channels,
whether or not these channels were viewed by the subscriber.
A second system was per-program pay TV, or a charge per
program watched.   This second system provided an electronic checking system to note if channels were in use by
subscribers.   Michigan State University personnel began to
develop equipment that would expand the then-current applications of two-way television for potential uses of interactive
television useful for all sectors of society.

The Michigan State University-Rockford project allowed
for the development of equipment with an interactive response
capability at the studio and terminal.   This required the
inclusion of a push-button type converter to transmit the
return signal from the subscriber (student).   In addition, the
minicomputer had to process response data in real time as
well as perform basic scan and maintenance routines.   Potential services gained by the development of this equipment
included in-home shopping; multiple-choice questions (and
answers) for a variety of programs (political polls, citizens'
response to community interests, response to video games,
educational programs); and simple fire alarm systems.

Based on the data gained in the development of this

second-generation response-television system, Michigan State University personnel developed a third-generation prototype. This equipment allowed more complete information to be received from the subscriber (student). More complex alarm systems indicating the need for medical help and invasion or burglar alarm signals were developed. This generation of equipment made possible electronic scanning for utility services in the private sector and scanning inventory needs for businesses.

Project investigation revealed that daily training is an on-going task in many government and social departments. Two-way cable television could provide many organizations with standardized, routine training updates and reviews. The Michigan State University-Rockford project focused upon the in-service needs of fire fighters. Prepared materials were sent via cable television to the fire stations or the Training Academy. Computerized scanning determined if the fire fighters were able to participate in the specific training unit, or if the fire fighters had responded to a fire alarm. (A unit was dropped from the feedback system when two consecutive questions were not answered by the participants.) The project included a series of interactive televised lessons concerned with pre-fire planning; nearly 200 fire fighters participated in the total study. Various methods of recording the data received during classes were used. In the time since this project was completed many of the recording methods have been shared with the business, educational, and home consumer sectors.

The Michigan State University-Rockford project also involved in-service programming for public school faculty in the Rockford area. This phase of the project was hampered by communication and scheduling difficulties. However, valuable data were obtained which were useful for in-service programs in succeeding years.

At the conclusion of this National Science Foundation project, the prototype equipment was dismantled and returned to the Michigan State University campus. Concepts and technical improvements developed during the course of the project have been shared with interested sectors of society. Cable companies in Portland, Oregon; Syracuse, New York; and Temple Terrace, Florida, are using the ICTV technology developed at Rockford. Nationwide, urban communities have noted the value of interactive television communication, and many cities are considering development of two-way cable

systems. As cable franchises are renewed, city administrators are considering the inclusion of two-way capability in
franchise specifications. It is estimated that by the late
1980's a large number of cities will have two-way cable television services developed as a result of the research and
knowledge gained in the Michigan State University-Rockford
study.

The educational tapes created for pre-fire planning
study by the fire fighters in Rockford are now used in various locations in Florida and Michigan in a one-way mode.
This provides improved learning modes for several groups
of fire fighters. The value of interactive television communication for education will be felt in many parts of society in
the coming years.

The improvement in equipment and in the technology
of two-way television gained through the Michigan State University-Rockford project study will have far-reaching benefits. Many services, such as invasion alarms and viewer-
response information, are utilized throughout the country
now. Other computer-response services will be available
for television viewers in the next few years, altering work
routines, educational procedures, consumer patterns, and
recreational/entertainment preferences.

REFERENCES

Baldwin, Thomas F.; Bradley S. Greenberg; Martin P.
    Block; John B. Eulenberg; and Thomas A. Muth. Mich-
    igan State University-Rockford Two-way Cable Project:
    System Design, Application Experiments and Public Pol-
    icy Issues, June 1978. Volume II. Department of
    Telecommunication, Michigan State University, East
    Lansing, MI 48823. (Final Report, NSF Grant No.
    APR75-14286.)

Baldwin, Thomas F.; Bradley S. Greenberg; Martin P.
    Block; John B. Eulenberg; and Thomas A. Muth. Mich-
    igan State University-Rockford Two-way Cable Project:
    Minicomputer System Software, 1978. Volume III.
    Department of Telecommunication, Michigan State Uni-
    versity, East Lansing, MI 48823. (NSF Grant No.
    APR75-14286.)

Baldwin, Thomas F.; Bradley S. Greenberg; Martin P.

Block; John B. Eulenberg; and Thomas A. Muth. Summary: Michigan State University-Rockford Two-way
Cable Project: System Design, Application Experiments
and Public Policy Issues, 1978. Volume I. Department
of Telecommunication, Michigan State University, East
Lansing, MI 48823. (NSF Grant No. APR75-14286.)

Baldwin, Thomas F., and others. "Experiments in Interactive Cable TV: Rockford, Illinois: Cognitive and Affec-
tive Outcomes," Journal of Communication, Autumn
1978. Annenburg School of Communication, 3620 Walnut
Street, Philadelphia, PA 19140.

Brownstein, Charles N. "Experiments in Interactive Cable
TV: Interactive Cable TV and Social Services," Journal
of Communication, Autumn 1978. Annenburg School of
Communication, 3620 Walnut Street, Philadelphia, PA
19140.

Clarke, Peter, and others. "Experiments in Interactive
Cable TV: Rockford, Illinois: In-Service Training for
Teachers," Journal of Communication, Autumn 1978,
Annenburg School of Communication, 3620 Walnut Street,
Philadelphia, PA 19140.

☐   The Rand Corporation, in cooperation with Telecable
Corporation and South Carolina public service agencies, submitted a successful proposal to the National Science Foundation in 1974.   The site selected for the proposed project
was Spartanburg, South Carolina, a community situated in the
northwest part of the state.   The population at the time of the
project was 45,000; 7,000 area subscribers received cable
television on twelve channels.

The purpose of the project was to study the public
service applications of two-way video systems.   The Spartanburg cable system contained a proven two-way video and
data-return cable system of high technical quality.

One experiment conducted within the total study examined the use of interactive television to educate day-care
directors and caregivers, and through this process, upgrade
the quality of day-care services.   One method provided workshop study with audio return and video capability, that is,
learning with an opportunity to interact instantly with the
teaching center and the other workshops through electronic
technology.   The second method provided these two-way workshops "live" for viewing by caregivers in the traditional one-
way television format.   Data gathered from these two experimental methods of study were compared with a control group.
The control group did not receive workshop study material in
any format.

Other primary investigation studies involved adults
using telephone and computer equipment for home study.
Course study involved completion of the high school equiva-
lency examination (GED), or a course in childrearing for
parents in the community.   A secondary investigation, growing out of the GED Study, provided a longitudinal study.
Data gathered in this secondary study were concerned with
timed interactions between teachers and students in both
traditional and interactive television classes.   All the studies
in this project provided valuable data for in-service training

programs and specific applications of interactive television
for local programming.

REFERENCES

Bazemore, Judith S. "Two-way TV Technology and the Teaching of Reading," Journal of Reading, 21(6), (March 1978):
518-24. International Reading Association, Inc. P. O.
Box 8139, Newark, DE 19711.

Berryman, Sue E.; Tora K. Bikson; and Judith S. Bazemore.
Cable, Two-way Video, and Educational Programming:
The Case of Daycare, October 1978. Rand Report Num-
ber R-2270-NSF, The Rand Corporation, 1700 Main
Street, Santa Monica, CA 90404.

Lucas, William A. "Experiments in Interactive Cable TV:
Spartanburg, S.C.: Testing the Effectiveness of Video,
Voice, and Data Feedback," Journal of Communication,
Autumn 1978. Annenburg School of Communication, 3620
Walnut Street, Philadelphia, PA 19140.

Lucas, William A. Moving from Two-way Cable Technology
to Educational Interaction, August 1976. Rand Paper
Number P-5704, The Rand Corporation, 1700 Main
Street, Santa Monica, CA 90404.

Lucas, William A. Two-way Cable Communications and the
Spartanburg Experiments, September 1975. Rand Paper
Number P-5484, The Rand Corporation, 1700 Main
Street, Santa Monica, CA 90404.

Lucas, William A.; Karen A. Heald; and Judith S. Bazemore.
The Spartanburg Interactive Cable Experiments in Home
Education, February 1979. Rand Report Number R-2271-
NSF, The Rand Corporation, 1700 Main Street, Santa
Monica, CA 90404.

Lucas, William A., and Suzanne S. Quick. Serial Experimentation for the Management and Evaluation of Communication Systems, September 1977. Rand Paper Number P-5989, The Rand Corporation, 1700 Main Street,
Santa Monica, CA 90404.

☐   The community of Reading, Pennsylvania, has probably had two-way television via cable longer than any other
community in the United States.   A locally-owned cable service, The Berks Cable Company, was started in Reading in
the early 1950's, located on the east side of the Schuylkill
River.   This cable company was purchased by Milton H.
Shapp, who, when he was elected Governor of Pennsylvania,
sold it to ATC (American Television and Communication
Corporation) in 1970.   Another cable system, the Suburban
Cable Company, on the west side of the Schuylkill River,
began during the 1960's.   ATC bought the Suburban Cable
Company in 1970, a few months after it bought the Berks
Cable System.   Thus, the two cable systems became the
Berks-Suburban Cable Company of ATC under one management.
It is now called the Berks Cable of ATC, Inc.

Rather than repair the worn cable found in these long-
existing systems, ATC officials elected to install a completely new system in the Reading area.   This situation
created an opportunity for two-way cable transmission.   (At
that stage of television transmission development, two sets
of cables were needed; one cable downstream, and another
upstream provided better transmission quality.)   And so,
experimental programs with two-way transmission were conducted over the next seven years.   These programs were of
public interest and were often concerned with local government issues.   One program, "This Is Your Mayor," began
in early 1967 and was televised directly from the mayor's
office.   At that time audio return, through telephone, provided live communication with the viewing audience.   The
League of Women Voters also began presenting information
to the public on a continuing basis.   Cable television subscribers in the Reading area began to consider two-way
television as an "ordinary" presentation of television programming, not realizing that other areas of the United States
did not have this format of programming available.

The Alternate Media Center of the School of the Arts

of New York University together with the City of Reading
applied for and received a National Science Foundation (NSF)
grant in 1974. This grant provided personnel and equipment
to expand the two-way television services. The focus of
this grant was to analyze the value of interactive television
for senior citizens in the cable-viewing area. Much information in the sociological and communication fields has developed as a result of this NSF study in Reading.

As part of the NSF study, three senior citizen centers were equipped with two-way television studios. The
assistance of senior citizens in operating these studios was
requested. As a result many senior citizens were trained
in the various functions of studio operation, and for many
years the centers have been staffed entirely by senior citizens.
Program content has included participation activities such as
"Sing-Along, " a weekly half-hour program with all senior
center audiences singing familiar songs of yesteryear and
today. Other programs focus on the information needs of
older people. Aspects of Social Security services, nutrition,
and other concerns of aging are presented with opportunities
for questions and discussion from the audience. Cultural
interests, demonstrations of craftmaking, cooking, gardening,
and leisure interests are all considered in-depth in special
presentation programs.

Another aspect of the NSF study provided direct audio
return from the homes of cable subscribers into the television
studio. The value of audio participation from residential
sites was tested in 135 subscriber homes. Response was so
favorable and so positive that, at the conclusion of the NSF
study, ATC provided audio return to the studio for all cable
subscribers. Remote audience participation has proved to be
a valuable, desired component of the Reading Berks County
TV cable system.

Interactive television in the Reading area is live,
simultaneous communication from person to person. Enthusiasm for and participation in the interactive programs during
the NSF study were so great that the nonprofit organization
(Berks County Television) that was developed during the NSF
grant continued after the end of the grant period. During
1982 new facilities were completed, and the BCTV studio is
thriving and growing during the 1980's. Many special audiences receive two-way programs through BCTV, including
local hospitals, businesses, industry, religious groups, city

and county governments, mental health organizations, and
cultural and educational institutions.

Berks Schoolcasting (Reading)

        Berks Schoolcasting was formed in September 1974 in
an effort to utilize a common cable-television channel efficiently through the presentation of school- and community-produced
programming.   Channel 5 was selected as the primary channel for this school-related material.

        Berks Schoolcasting is a voluntary organization of
media personnel from local industry, government, and school
institutions.   Membership is open to all public, parochial,
and private schools and colleges in Berks County.   No mone-
tary dues are required, only a desire to work on a cooperative basis.   Individuals meet monthly during the school year
to coordinate the schedules of the locally-produced television
programming.

        The organization of Berks Schoolcasting has a history
dating back through the 1960's.   In a desire to determine
efficiently a feasible system of providing educational television to the Berks County schools and viewing community,
various committees were formed.   These committees explored ways in which educational television might be provided to school and community members, analyzed cost
factors for each proposed transmission system, and considered the overall possibilities for educational television
services.   During the long process of examining several
systems, one pilot educational program emerged and was
adopted.   A cooperative arrangement between the schools and
the cable company developed.   The schools purchased and
installed the necessary equipment, and the cable company
wired the buildings and interconnected the schools through
the cable system.   A control center was established through
which classes were able to receive educational and commercial broadcasting as well as in-house instructional programming.   This plan was approved and implemented in the fall
of 1969 by the Berks County Television Advisory Committee
and has continued as the major transmission connection for
the Berks Schoolcasting program.

        During the early television experiments two schools,
Governor Mifflin and Wilson, almost accidentally stumbled
upon two-way interaction.   Each school had been given a
mid-band channel; and classes found they could share these

channels, and also see and talk with each other simultaneously. Thus, two-way television classes began.

Early in the 1970's, seven major organizations in
Berks County banded together in order to develop an area
cable/microwave system: the Alternate Media Center of New
York University; Berks Intermediate Unit 14; Berks-Suburban
Cable Company of America Television and Communications
Corporation; Keystone Cable Company; Kutztown State College;
Publi-Cable Inc.; and Schuylkill Intermediate Unit 29. A
series of meetings were held, and a regional Publi-Cable
Association was formed in Pennsylvania. Membership grew,
organizational direction shifted, and a need for in-service
training of teachers in the use of television as an interactive
medium was determined. A three-year grant from ESEA
(Elementary and Secondary Education Act) Title III provided
for the development of ITEL (Instructional Television for
Experiential Learning).

The Schuylkill Intermediate Unit 29, with the endorse-
ment of the Berks County Intermediate Unit 14, served as Local
Education Agency for the ITEL project. However, the project was subcontracted to and conducted by KSC-TV Services
of Kutztown State College. ITEL had three major purposes:
1) to instruct educators in the use of one camera/one VTR
operation in generating curriculum content; 2) to assist
educators in instructing colleagues in the home schools with
the interactive use of television; and 3) to utilize and share
cable television materials whenever and wherever possible.
Two major long-range actions resulted from the ITEL project. The first was a growing involvement with the National
Publi-Cable Association; Kutztown State College was selected
as the site for the second National Publi-Cable Conference
in April 1972. KSC-TV Services of KSC and Berks-Suburban
Television served as co-hosts. Clay Whitehead, then head
of the White House Office of Telecommunications Policy,
gave the keynote address and interacted, via television, with
Mayor Shirk of Reading. This occasion marked the first use
of microwave as an interconnect with cable in Berks County.

The second major long-range action resulting from
ITEL encouraged the development of television facilities in
area school districts. Thus, schools became involved in
using the Berks Cable system as an institutional network.
Through this growing interest in the educational uses of
cable television, ITEL became one of the contributing causes
for the creation of the Berks Schoolcasting organization.

In 1974, the National Science Foundation study developed. Expansion of television transmission and equipment followed rapidly, and school districts shared in the improved transmission. By the end of the 1970's, schools in Berks County had developed much variety in educational television utilization. School districts tended to develop areas of specialization, which have been willingly shared with all members of Berks Schoolcasting.

A long-standing administrative educational unit, the Berks County Intermediate Unit (BCIU), developed a film library. Using the cable television system, films were aired to schools for educational purposes. Today, BCIU continues to highlight downstreaming of films and videotapes to institutions that are in the system. This system is a closed-circuit channel transmitting only to members of Schoolcasting. Materials shown have been cleared for viewing by an extended educational audience prior to transmission. Each school determines which films to air in the individual district. This BCIU program is the only use of pre-recorded materials within the Berks County/Schoolcasting television system. (Other transmission materials are live, interactive programs.) However, the beneficial value of recorded films to enlarge learning experiences allows this phase of television programming to continue. There are over 400 films in the BCIU library, and nominal fees are charged for transmission. Schools not on the cable pay a necessary fee to dub copies for later broadcast. The Berks County Intermediate Unit also serves as a scheduling center for television programming. Two channels have been activated for educational television programs, and the committee meets once a month to determine scheduling for all school districts.

The Conrad Weiser Area School District has been involved in limited television production and utilization since 1970. In 1980, a $50,000 school television studio and distribution facility was completed, and a full-time staff member added to manage it. In conjunction with the Wernersville State Hospital and the local chapter of the Future Nurses Association, students at Conrad Weiser have developed special television programs. The content of the programs focuses on the needs of patients and is useful for instructors wanting to use video in patient education.

Governor Mifflin School District began monthly cable cast production in October 1972. Since then programming has steadily increased. Local origination is developed through the school studio: GMTV-Studio 713. From GMTV, sports events from the schools, community programming involving organizations such as Goodwill Industries, the League of Women Voters, and local dancing teachers have been televised to the schools and cable community.

Governor Mifflin operates a closed-circuit television system of six channels (regular and mid-band) over three individual distribution tracks. From the central control area, television programming from BCIU, the GMTV studio, or state-produced instructional materials may be transmitted. Interest and expansion of television production in the system has been encouraged by regular student summer workshops and training sessions; student operation of equipment and participation in television activities is fostered and maintained. Programming is produced in the district at all levels from kindergarten through secondary classes. In 1979, the Governor Mifflin School District received the Philadelphia Award from the Classical Society in Philadelphia for developing a series of programs in Latin. Access to television programming for Governor Mifflin Schools can occur in many different procedures (see Chart I on pages 44-45). This variety in television reception/transmission is true for all schools in the Berks County area.

The Reading School District began television programming in 1968 with "Knight Life," a program produced weekly, highlighting activities of the school. Participation is an extra-curricular activity for about 25 students; the programs are aired Monday evenings throughout the school year. "Knight Life" is the longest continuously-running program of any kind in the Berks Cable system. "Knight Life" has two major objectives: first, to explain to the public what Reading School District students are accomplishing, not only within school activities, but also in the community at large; second, to provide students interested in the field of communications with an opportunity to explore the world of television, either before or behind the cameras, and to develop their skills in the field. Reading School District students originated the interactive programs with the mayor and city council members in 1974 and were involved with the critical issues programs shared with the senior citizen audiences during the NSF study. Newer developments include organization of the Appalachian Community Satellite Network programs to use television for many continuing education courses and for the GED. Using an 800 telephone number, students can interact through two-way audio with the teachers for these adult courses.

Chart I:

ACCESS CAPABILITIES: GOVERNOR MIFFLIN SCHOOLS

● Intradistrict programs for use within one classroom, several classrooms, or several buildings
Accomplished through one public channel, or one private (mid-band) channel utilized by the school alone.

● Interdistrict programs for simultaneous use by more than one school district
Utilizes switching equipment owned by the cable company so that home viewers can see interschool programs; also, two-district interactive programming can occur on mid-band channels.

● School District News
A news show, "Fanfare," has been produced since October 1973. The programs are broadcast twice on alternate Wednesdays and Fridays; each show mentions events and activities for the coming two-week period. Junior and Senior High students participate in the productions of this show.

● District Feature Program
For the past ten years the school system has produced a weekly program designed to examine some aspect of the total program operated by the school system.

● Sports Program
During the winter season, as a public service, specific athletic contests are broadcast live. Athletic contests of Junior and Senior High girls and boys are featured.

● Cooperation with Community Groups
The school district has handled program preparation and cablecasts in cooperation with Goodwill Industries, a local Lions Club, the Shillington Fire Company, the League of Women Voters, Reading Boys' Home, and the Governor Mifflin Community Days committee.

Chart I (cont.)

- In-service Programs
  Sample teacher lessons have been replayed for study
  by several other teachers in other buildings.

- Special Seasonal Programs
  School opening and closing, Christmas, Halloween
  and other special activities have been cablecast
  through the years.

- Special Events Programs
  Groundbreakings, individual school activities, gym-
  nastics programs, assemblies, visiting personalities,
  School Board activities and similar events have been
  videotaped and later cablecast.

- Berks Schoolcasting
  Much public-access programming on a monthly basis
  is cooperatively scheduled with the cable company and
  the other members of Berks Schoolcasting.

- Berks I. U. 14 I. M. S. Motion Pictures via Channel J
  School districts request video-available (secured legal
  rights) materials to be transmitted system-wide from
  the film and videotape library of the Berks Instructional Materials Service. Scheduling requests on planned
  on a monthly basis.

This chart has been adapted from BERKSchoolcasting TV 5.

Community service programs have also continued to be important to the television production at Reading School District.

Wilson School District pioneered in cablecasting to the
community in 1969. Football games were provided via delayed cablecast, and school wrestling matches and basketball
games were transmitted "live." Wilson is the only school
district in Berks County in which the monthly school board
meetings are cablecast "live." Reoccurring programs have
included music concerts and Special Olympics events.

Over 50 students help produce the "Wilson Highlights"
each Monday evening. Two complete crews, each with 25
students, produce the hour-long program, alternating weekly
programming schedules. Each crew includes a technical
staff and a news staff. Television production began with only
one camera, but has expanded to a four-camera studio and a
separate control room area. The in-house closed circuit
system is sent to all buildings via a five-channel system.
Playback of locally produced programming is extensive, and
this form of transmission continues to increase. In the
school year 1971-72 about 650 hours of playback programming occurred. Today, over 4,000 hours of playback time per
school year takes place. A mass media course, the English
Department, and American History courses have all included
a two-week mini-workshop in television production. Students
do the actual production and content is related to the disci-
pline of the course. Many additional classes in the school
utilize television production at some point during the semester. Students gain from experience in cooperating with others
while working as a team member, studying/researching
course content, learning in a new and more enjoyable manner, and absorbing knowledge about a new field of communication and technology.

Kutztown Area School District, located north of Reading, uses off-air videotape production, programming from
BCIU, and live television. Kutztown students have presented
a weekly community program in which viewers are asked to
telephone the studio and interact with students on the program.
Issues of current interest to citizens and students are discussed; guest experts are invited to the school, and advance
notice of the topic for the week is promoted through regional
media. Production of in-house television is an extracurric-
ular opportunity for students in Kutztown schools.

Wyomissing Area School District began television use

in 1972 by wiring buildings for internal distribution of cur-
ricular programs.   No cable system was able to provide
services to the school until 1974.   At that time, Berks Cable
of ATC activated a cable system that linked four of the five
Wyomissing school buildings together.   This installation also
allowed connection to the Berks Cable System and hence, to
the other school districts participating in the interactive
system.   Interdistrict programming became a reality for the
Wyomissing schools.

One of the highlights of the Wyomissing Area School District curriculum is computer instruction via interactive television.   Instructors in the high school mathematics department
send information downstream via the two-way television system,
through interconnected terminals and a series of videotapes, to
the mathematics department in the middle school.   Audio talkback by telephone is "live," with sound amplification.

Central Catholic High School has the distinction of being the first school to cablecast local programming.   The
school has been involved with television and the Berks Suburban TV Cable Company since 1967.   Central Catholic has
its own television production studio for live or pretaped programming.   Program content centers around groups and activities of the school.   During the NSF study the school was involved in an oral history program in which students interacted with senior citizens at the senior centers through the
two-way television system.   Although the NSF grant has long
since terminated, the oral history program continues and has
expanded to include several high schools in the area.   Additional uses of dialogue through the interactive system have
included conversations between political candidates at the
state and national levels and students at Central Catholic
High School.   This type of programming has been modified
to include local and regional candidates, members of the
senior citizens groups, and the Chamber of Commerce.   All
programming provides enrichment and increased awareness
of social and cultural concerns in the Berks County area.

Kutztown State College (KSC), mentioned earlier, continued to expand the use of cable and microwave link television
long after the conclusion of the ITEL project.   By the fall
of 1976, KSC conducted a pilot project with two-way television
through teaching a course in Oceanography.   Students from
three area high schools interacted via microwave/cable link
with Kutztown and Berks Community Cable Companies.   This
project, plus the many demonstration programs conducted

from 1972 through 1976, proved that the microwave/cable
television system in Berks County has the necessary capabilities to teach college level courses through television to
several locations simultaneously.

In 1977 KSC received a grant from the National En-
dowment for the Humanities allowing expansion beyond the
pilot project concerned with Oceanography. This grant facil-
itated the development of two language courses via interactive
television. Students in five area high schools studied French
and Russian with KSC professors. Data from the study indicate that students learned at a rate comparable with the college students; all students were evaluated on the same basis.
The many projects in the KSC cable/microwave television
programming clearly indicated that the technology of interactive television has an educational/informational potential
limited only by the creativity and interest of the individuals
involved in the television systems.

Nearly ten years ago influential people at the Reading
Hospital and Medical Center wanted help with operating techniques practiced in Dallas. Via special interactive television
transmission, an opportunity for observing and learning this
surgery was created. From this special learning situation,
transmission systems for continuing medical educational programming developed. Community Medical Cablecasting, in
conjunction with the Berks County Medical Society and the
New York-based Medical Cablecasting, provides medical educational programs that are transmitted directly into the offices
or homes of participating physicians. A special channel,
requiring converters to unscramble the mid-band channel,
limits reception to the designated audience. Cablecasts are
sent four times weekly over the Berks system.

In another related medical service, Berks TV Cable
was asked to plan and produce a film documentary concerning the common tick, an insect that is the carrier of Rocky
Mountain Spotted Fever. With the assistance of the Department of Health of the Commonwealth of Pennsylvania, the
production has been developed and aired nationwide.

The local family guidance center considers interactive
television very useful as an observational tool and consultation service. In general, interactive television is used
within the Schoolcasting member hospitals in many ways. Assisting in consultation by distance is a special aid for staff
members, who often have very heavy schedules. Videotapes

developed for patient information concerning illnesses and for
self-help/self-care education are heavily utilized in the Berks
County medical facilities. There is some concern that a
depersonalized attitude will develop when equipment occasion-
ally replaces face-to-face interaction. However, live two-
way television does provide many advantages and opportunities
for communication that without electronic means would be
impossible. Long-range decisions regarding interactive television use remain undetermined at this point.

Wernersville State Hospital, southwest of Reading,
was awarded a grant from the Bureau of Telecommunications
Demonstration Grant Projects of HEW in the fall of 1977.
This grant allowed Wernersville State Hospital to link up
with the existing two-way microwave cable network. This
transmission link has permitted the hospital to expand its
program to return patients to the mainstream of society by
interfacing with the various types of two-way television programs aired through BCTV. "Sing-Along" is well received
by the patients at Wernersville. Programs from the Appalachian Satellite Community Network are useful and inform-
ative for some of the patients. In addition, two programs
originated at Wernersville are shown daily, and there is also
a locally-produced weekly series. Staff members are gradually increasing their use of two-way television for therapy.
Pastoral services and patient self-analysis use two-way television. The patients observe videotaped sessions of their
interactions with staff during the counseling sessions and
self-analysis of their specific situation occurs through this
viewing process.

The series of meetings which began in the late 1960's
"to efficiently determine a feasible system of providing educational television to the Berks County Schools" have resulted
in an organized, thoughtful approach to the application and
utilization of two-way television. The members of Berks
Schoolcasting, as well as all the organizations that have provided support and guidance through the years, can look with
pride to the cable/microwave television system in Berks
County. Attention to program content, in quality and variety,
has increased, and audience interest and enthusiasm have
been maintained. Attention to technology, in providing clarity in transmission signals, flexibility in program origination,
and opportunities for live interaction, has provided Berks
County citizens with electronic communication of a kind not
found in many other places in the world. Berks County students are receiving training that will equip them for employment

in technology developments during the coming years. Berks
County citizens are able, through two-way television, to
stay informed of local activities and information, enriching
the quality of daily life. The Berks County area cable/micro-
wave system has provided an example and can serve as a "pilot program" for other communities throughout the world.

REFERENCES

Award Winning Local Television Production, 1970. Berks
    TV Cable Company, American Television and Commun-
    ications Corporation. Berks Cable Company, 400 River-
    front Drive, Reading, PA 19602. (215-378-4600)

BERKSchoolcasting TV 5, published by members of Berks
    Schoolcasting Organization. Contact members of Berks
    Schoolcasting or Dr. Robert Fina, Kutztown State College,
    Kutztown, PA 19530.

Brumbaugh, Robert; Roberta Crisson; R. P. Fina; and Jacob
    Mandel. "ITEL," ITEL: Instructional Television
    Through Experiental Learning, 1975. Kutztown State
    College, ESEA Title III, Kutztown, PA 19530.

Burns, Red, and Lynne Elton. "Experiments in Interactive
    Cable TV: Reading, Pa.: Programming for the Future,"
    Journal of Communication, August 1978. Annenburg
    School of Communication, 3620 Walnut Street, Philadel-
    phia, PA 19140.

Connell, Eileen. "Experiments in Interactive Cable TV:
    Reading, Pa.: Training Local People," Journal of Com-
    munication, Autumn 1978. Annenburg School of Commun-
    ication, 3620 Walnut Street, Philadelphia, PA 19140.

Fina, Robert Patrick. SCCITEL: Satellite/Cable Community
    Interactive Telecommunications, 1982. Berks Intermedi-
    ate Unit Number 14, Antietam Valley Shopping Center,
    Reading, PA 19602. (215-779-7111)

Fina, Robert. "Two-way TV in Kutztown," Videography,
    April 1979, pp. 24-5. United Business Publications,
    475 Park Avenue South, New York, NY 10016.

Moss, Mitchell L. "Experiments in Interactive Cable TV:
    Reading, Pa.: Research on Community Uses," Journal

of Communication, Autumn 1978. Annenberg School of Communication, 3620 Walnut Street, Philadelphia, PA 19140.

Moss, Mitchell L. Summary: Two-way Cable Television: An Evaluation of Community Uses in Reading, Pennsylvania. Final Report to the National Science Foundation (Grant Number APR75-14311 A02). The NYU-Reading Consortium (New York University, Alternate Media Center/School of the Arts, Graduate School of Public Administration, April 1978), 144 Bleecker Street, New York, NY 10012.

Moss, Mitchell L. Two-Way Cable Television: An Evaluation of Community Uses in Reading, Pennsylvania. Volumes I and II. Final Report to the National Science Foundation (Grant Number APR75-14311 A02). The NYU-Reading Consortium (New York University, Alternate Media Center/School of the Arts, Graduate School of Public Administration, April 1978), 144 Bleecker Street, New York, NY 10012.

The Reading Dialogue. American Television and Communications Corporation, n.d. Contact: Berks Cable Company, 400 Riverfront Drive, Reading, PA 19602. (215-378-4600)

Richter, Jerry. "Berks Community TV's Interactive Ability Is Integral Part of Reading City Government," Community Television Review, 4(3), (July 1981):8+. University Community Video, 425 Ontario SE, Minneapolis, MN 55414.

Richter, Jerry. "Interaction: The Misuse of Term 'Interactive,' Ability of People to Talk to People Should Be Assured in Cable Design," Community Television Review, 4(3), (July 1981):34+. University Community Video, 425 Ontario SE, Minneapolis, MN 55414.

IX. CALIFORNIA STATE UNIVERSITY AND COLLEGES
TELECOMMUNICATIONS NETWORK

☐   Over the past decade the California State University
and Colleges have been developing a telecommunications network to serve academic and administrative needs. In 1979
a five-year plan was developed that will provide expanded
services through microwave and ITFS transmission. Video,
audio, data, and facsimile (including electronic mail and
library automation) will provide complete communication
between campuses. Projected estimates indicate that financial expenditures will be cost effective, efficient, and reliable. In addition, development of a Telecommunications Network owned by CSUC will allow for the implementation of
communication transmission specific to CSUC needs, with
flexibility in development to meet future requirements.

The CSUC system has nineteen campuses extending
from Chico in northern California to San Diego at the southern edge of the state. Over the years, a "backbone" communication network has been established among some of the
CSUC locations. All nineteen campuses, plus communication
services to the Chancellor's office and the Division of Information Systems (DIS), will be in the network by 1985. CSUC
locations that have implemented interactive television classes
(primarily one-way video, two-way audio) have had a positive
response from students and faculty, and course offerings and
enrollment are expanding rapidly.

Communication services provided in this project include the following:

> video classes -- two-way audio and one-way video
> sent to multiple off-campus sites, with "live"
> audio return. Course content includes degree
> programs, and educational extension

> telephone conferencing, satellite, or microwave
> "talkback" telephone network between campuses

> data communications -- including the communication

technology for time-sharing computers, providing maximum efficiency

facsimile -- electronic mail, library automation, and other types of printed transmission, including slow-scan television.

These services will interconnect all 19 campuses, providing increased efficiency and economy in communication and resource sharing. By developing an electronic system under CSUC ownership, equipment development can be adapted as future electronic requirements are determined, including the use of satellites for state-wide live instruction. In the more distant future, a second channel may be added to the backbone system if needed and feasible. At that point, increased use of two-way video transmission could be implemented if desired.

The California Statewide System has recently contracted with Control Data Corporation to market the PLATO system throughout California and the Western states. The intended audience for PLATO materials will be all public service organizations and educational institutions from kindergarten to university level.

The expansion of many technological services is on the horizon in California. San Francisco State College is experimenting with the use of videotext for information sharing. Soon, computer terminals will be able to search university library collections in moments and provide twenty-four- to forty-eight hour delivery of resource materials to students. Facsimile services for needed course materials are increasing, as are microfiche collections. However, even with the advances in technology individual advising and counseling continues to be important in California schools. Technology increases access to information and helps to speed and clarify the learning process. Technology will be installed to assist the student in finding greater uses for information and knowledge and to enhance the quality of living for each person.

REFERENCES

Loughboro, J. Patrick, and Dick Hampikian. A Telecommunications Network for the California State University and Colleges, June 1979. JP Associates, Inc. 3115 Kashiwa Street, Torrance, CA 90505. (213-539-8533)

X. STANFORD UNIVERSITY

☐   In 1953 the School of Engineering, Stanford University, California, developed an Honors Cooperative Program (HCP) to assist qualified engineers and scientists from nearby businesses to obtain graduate degrees. The students were professionally employed full-time, so graduate study was pursued on a part-time basis. HCP students had the same rights, privileges, and responsibilities as on-campus students. Through the years, in expanding the HCP program, the Stanford Instructional Television Network (SITN) was developed. This instructional network now provides interactive educational television with one-way video and two-way audio broadcast for twelve hours a day, over four channels allocated by the Federal Communications Commission. This interactive television system can be received at company sites within a 50-mile broadcast range of Stanford. Signals are transmitted via a microwave link to a transmitter located on Black Mountain, ten miles from campus. In turn, ITFS signals are sent to the designated sites, then converted to a standard VHS television channel.

Stanford University provides technical specifications and guidelines for equipment, makes available a service contract to service equipment, and encourages educational experiences of the highest quality. Each participating firm is responsible for classroom space and equipment. In addition, each company assigns a coordinator to process student enrollments and take care of on-going program details in the receiving classroom. Over time, the program has expanded to serve several thousand professional engineers, managers, and scientists from 100 San Francisco Bay area firms and research institutions, all members of SITN.

Organizations interested in SITN assume three financial obligations when considering long-term membership. First is a membership fee to help with the cost of the interactive television system. A trial membership of one year is possible before assuming a long-term commitment. The second obligation is the purchase of receiving or playback

equipment for company classrooms. This cost depends upon
the inclusion of talkback facilities, and the number and design
of the classrooms. The third obligation is that SITN members
assume responsibility for student fees. Firms agree to participate in the Honors Cooperative Program. Tuition is
charged on a per-unit basis, not to exceed one-fifteenth of
the total quarterly tuition charge for full-time students.
There is also a per-unit matching fee charge equal to the
tuition, plus a per-unit television charge. Taping charges
are additional.

The two major components of the instructional television network are the School of Engineering, Stanford University, and the Association for Continuing Education (ACE).
SITN provides over 150 Stanford courses per year in applied
physics, math, statistics, computer science, and all fields
of engineering. SITN also provides the Tutored Videotape
Instruction (TVI) program. ACE provides more than 80 non-
Stanford courses in business administration, management
development, supervision, and general interest subjects.
Through the Golden Gate University, member firms of ACE
may send students to Golden Gate University via the SITN
system. These students, all employed full-time, may receive graduate course work leading to an MBA degree from
Golden Gate University or undergraduate courses in business
through the College of Notre Dame.

In 1973 an extension of the program was developed:
the Tutored Videotape Instruction program (TVI). Originally,
the program began in order to provide engineering courses
to students employed at the Hewlett-Packard plant in Santa
Rosa, California, 100 miles north of Stanford. However,
high quality microwave transmission could not exceed 80
miles, so an alternate method was developed. Interactive
television classes were videotaped, and the videotapes were
sent to Santa Rosa. In Santa Rosa, tutors provided by the
company and approved by Stanford University viewed the videotaped classes with the students. The tape could be stopped
for discussion, questions, and interactions among the students
and tutor at the remote site.

This method of presentation/teaching was well received
by the students, who were expected to complete all assignments, as were the on-campus and other off-campus students.
The academic performance of students was very satisfactory,
and the program was expanded the following year to include
Hewlett-Packard plants at San Diego and Boise, California,

and the Sandia plant in Albuquerque, New Mexico. The Tutor
Videotape Instruction program is now employed at over 25
locations beyond broadcast range.

REFERENCES

Association for Continuing Education. MBA Degree; Golden
    Gate University, 1980 (revision). Golden Gate Univers-
    ity, 536 Mission Street, San Francisco, CA 94105.

Gibbons, J. F.; W. R. Kincheloe; and K. S. Down. "Tutored
    Videotape Instruction: A New Use of Electronic Media
    in Education," Science (Reprint Series), 195 (March 18,
    1977):1139-1146. American Association for the Advance-
    ment of Science, 1515 Massachusetts Avenue, N. W.,
    Washington, DC 20005.

Schedule of Classes, Summer 1981-82. Stanford Instructional
    Television Network, School of Engineering, Stanford Uni-
    versity, 345 Durand Building, Stanford, CA 94305.

XI. UNIVERSITY OF SOUTHERN CALIFORNIA, LOS ANGELES

☐ The University of Southern California, Los Angeles, has an interactive television network similar to the network at Stanford University. Four interactive channels are utilized, transmitting courses to 27 company sites in the greater Los Angeles area. A new relay, completed in early 1982, extends transmission to northern and central Orange County. Over 100 engineering, computer science, and mathematics courses are offered for credit. Approximately 50 noncredit courses are also provided via interactive television.

Over 4,000 students participate in the educational television courses. Programming originates in the Norman Topping Instructional Television Center, located in the Olin Hall of Engineering at the University of Southern California. Four studio classrooms are employed, broadcasting live from early morning until 9:30 p.m. Each classroom has three black-and-white television cameras. One camera is positioned over the instructor's desk, to zoom in on display materials, demonstrations, and television notes. Another camera positioned at the rear of the room focuses on the instructor and his or her on-going presentation. The third camera televises the students in the originating classroom for viewing by off-campus students. The Center also has a mobile television classroom which may be driven to company locations for live demonstrations.

Transmission is in the Instructional Television Fixed Service range, 2500 to 2690 MHz. Signals are relayed to a transmitter in Hollywood Hills, and then rebroadcast in all directions throughout Los Angeles, to a range of approximately 30 miles. Signal adapters convert the ITFS signals to frequencies that can be received by company television monitors. A relay station built on Signal Hill broadcasts programs into Orange County. Return audio is either by telephone or by FM radio.

As with the Stanford Instructional Television Network,

employers generally pay all or part of the tuition costs for
their employees.   With the University of Southern California
network, there are fixed monthly charges.   USC provides
companies with the paraboloid receiving antenna and the television downconverter, as well as routine maintenance and
courier service for instructional materials.   Participating
companies provide the television monitors and the audio
transmitters or telephones needed for student response.
Individual companies also determine the number of rooms
reserved for ITFS reception.   If desired, four channels can
be received simultaneously, or one classroom can receive
any of the four channels at various times.

REFERENCES

Munushian, Jack.   "USC Plans Major ITFS Expansion," The
    EMCEE Roundtable: ITFS in America Today, 1, (1),
    (February 1982):11-13.   EMCEE Broadcast Products,
    Inc.   P. O. Box 68, White Haven, PA   18661.
    (800-233-6193)

Munushian, Jack.   Personal interview.   Professor of Elec-
    trical Engineering and Director of Instructional Tele-
    vision Network, School of Engineering, University of
    Southern California, Los Angeles, CA   90089.

USC Instructional Television Network, February 1982.   Uni-
    versity of California Systemwide Administration, Office
    of Information Systems and Computing, Berkeley, CA
    94720.

USC I-ITV, no date.   Contact the Interactive Instructional
    Television Program, Seaver Science Center 510, School
    of Engineering, University of Southern California, Uni-
    versity Park, Los Angeles, CA   90007.

XII.   UNIVERSITY OF CALIFORNIA, DAVIS

☐   The University of California, Davis campus has developed interactive television through microwave and ITFS technology.   Proposed development will place the initial thrust
of expansion on voice and data transmission.   At present,
courses from the Engineering Department at UC, Davis are
sent to the Livermore Laboratory nuclear research center.
Full-time employees are given released time to attend the
television courses, held at the Livermore Laboratory; students are able to obtain course credit leading to graduate
degrees in Engineering from UC, Davis.   Students also receive, via microwave transmission, engineering courses
from Stanford University and University of California, Berkeley.   Transmission includes live television coverage of the
professor, with live audio return from remote sites to the
on-campus studio classroom.   ITFS signals arc sent to
Roseville, California, to the Hewlett-Packard Plant from UC,
Davis for a similar type of study program.   The Hewlett-
Packard employees also receive course work via television
from the California State University, Chico.

One means of communication between student and professor at UC, Davis is affectionately dubbed "the confessional
hour."   Students wanting clarification in course material
schedule appointments and meet with the professor via return
video and audio.   A reverse process occurs in that the camera focuses on the student in the Livermore Laboratory classroom, and the student is "on camera."   The professor receives video and audio signals, and responds only through
the audio channel.   Response to this conference system has
been good.   In addition, faculty members are consequently
reminded of the technology signals sent to all students in the
course and are able to clarify all information sent to students
as the course progresses.   Courses, through technology, may
be sent great distances in California.   In 1982, UC, Davis
sent courses to Palo Alto in southern California, and to
Yreka, 285 miles north, utilizing the "backbone" communication system of the state.

One phase of interactive development has commenced in the School of Medicine, UC, Davis. An Instructional Television microwave system now links campus lecture halls with the UC, Davis Medical Center Auditorium, a distance of 18 miles. Lectures, symposia, and research seminars may be held at either site and transmitted "live" to the other locations. Also, operating, patient examining, and autopsy rooms are outfitted with video and audio equipment so that first- and second-year students may view these procedures via television. In addition, these sessions may be recorded for future replay at times convenient for students.

An increased use of video at UC, Davis began in 1981. Teleconferencing has proved to be economically viable, saving hours of commuting time for students and staff. The value of television transmission of microscopic materials is increasingly recognized. Electronic pointers, split screen techniques, and character generation information aid the teaching and learning processes. UC, Davis is providing state-of-the-art equipment to encourage newer methods of learning utilizing technology. In general, the staff of UC, Davis is pleased with interactive television developments and plans expansion of the system over the next few years.

REFERENCES

"Instructional TV Uses Expanded," UCDx School of Medicine, 1(2), May 1, 1981:5. School of Medicine, University of California, Davis. Contact: Public Information Office, School of Medicine, University of California, Davis Campus, Davis, CA 95616.

"Video Technology Used for a Geographically Split Campus," University: A Publication for Faculty and Staff of the University of Davis, 1981. Contact: Dean's Office, School of Medicine, University of California, Davis Campus, Davis, CA 95616

XIII. UNIVERSITY OF CALIFORNIA, BERKELEY

☐   In 1973, the University of California, Berkeley moved toward interactive television instruction by activating two ITFS channels on campus. The primary purpose was to provide upper-division and graduate courses leading to a Master of Arts in four disciplines within the School of Engineering. Receiving sites for the interactive courses are electronic businesses in the San Francisco Bay area.

An interactive television channel with Stanford University may be the most actively used ITFS channel. Seminars with Stanford University have included Slavic languages, exotic languages, and Chinese Studies. In early 1982, a Latin American Studies course focused on economic and social developments in Mexico. Classes were conducted at Stanford and at Berkeley, with professors at each location. Interactive audio and video, in color, aided in sharing knowledge and understanding. Students felt actively involved with learning while using this form of television. Classes in other subject areas, such as political science and Russian history, have been transmitted via ITFS. UC, Berkeley administrators realize that area studies need specialized faculty. Interactive television allows this specialized staff to be shared with larger numbers of students, based in many locations.

There is also a two-way television capability between UC, Berkeley and UC, Davis for interactive classes. In addition, data transmission between libraries has been heavily explored. Teleconferencing, computer searching, and display of reference and source materials have developed, with training and demonstration sessions for the librarians, using the two-way system.

UC, Berkeley currently has a project under way exploring the uses of interactive videodisc instruction. Five disciplines are developing instructional materials using still-frame visuals that will be transferred to videodisc. Students will be able to study materials through random access selection utilizing Apple II computers. Archaeology, English Lit-

erature, and Architecture have subject materials which will
transfer readily to the videodisc/random access format. UC,
Berkeley is also developing audiovisual materials for hearing-
impaired students, providing instruction in Sign Language. In
the field of Chemistry, a periodic chart combining videodisc
and off-line computer will create individualized random access
searching and information retrieval for students. This project is developing through the support of the Lawrence Hall
of Science, located on the Berkeley Campus.

        Another project which holds promise for the UC, Berke-
ley area is a cable television program entitled "Open Window. " Programming began in October 1981 with a two-hour
weekly time schedule. Content has focused on the educational
and cultural experiences of the UC, Berkeley Campus to the
community, airing student-produced materials, drama, dance,
interviews of important campus visitors, and thesis documen-
taries. Long-range plans for "Open Window" include developing a technical capability for interactive audio, so that cable
subscribers can interact during the program presentation, gaining information about the University and participating in current activities.

        A regional proposal for a higher education television
consortium for the San Francisco Bay area, using an acronym
of HITEL, has been submitted to the Annenberg Foundation
and to the National Telecommunications Information Administration (NITA). If funded, this grant would involve five
institutions, both public and private, in a three-county region
of San Francisco Bay. Junior colleges, state colleges and
universities are included in the proposed consortium. The
project is designed to create interactive television linkage
with the consortium institutions and the cable franchises in
order to develop telecourses and other television modules
across institutional lines and across higher educational levels.
Participating consortium classes would utilize portions of the
telecourse, as applicable for the specific course and level of
instruction. The proposed grant allows the HITEL Consortium
to serve as a model site, evaluating telecourse materials
from other areas. Publication of pedagogical intent, objectives, intended audience, anticipated outcomes, and mechanical problems would be one component of the HITEL Project.
Another desired component is involvement in the development
and production of the telecourse materials created for some
of the educational programs.

REFERENCES

"Open Window" on Satellite: The Next Stage in the University's Educational Use of Cable TV Networks, April 28, 1982. Opportunities for National Program Development and Distribution via Satellite, Educational Television Office. Contact Educational Television Office, University of California, Berkeley Campus, Berkeley, CA 94720.

"Video Disc: A Pilot Demonstration Project Illustrating Instructional Applications from Various Disciplines" (Application for Instructional Improvement Grant), 1981-82. Educational Television Office, University of California, Berkeley Campus, Berkeley, CA 94720.

Interviews with Peter C. Kerner, Coordinator, Educational Television Office, University of California, Berkeley Campus, Berkeley, CA 94720.

XIV. CALIFORNIA STATE UNIVERSITY, CHICO

☐   Twelve counties in northeast California, composing
one-fifth of the state's area (about 33,000 square miles),
utilize some of the newer telecommunication facilities emerging in the 1980's.   Mountains, weather, small communities,
and vast amounts of unsettled land contribute to a sense of
rural isolation for the 600,000 inhabitants.   In 1972 the
California Coordinating Council on Higher Education (CCHE)
released the results of an extensive needs assessment study.
This report indicated that large numbers of citizens in the
northeastern area of California were potential students for
continuing education in many subjects areas.   The need for
providing, and delivering, quality educational services to the
inhabitants of this area triggered the development of the interactive television system in northeastern California.

    In 1975 the system began operation.   Transmission is
in the 2500- to 2690-MHz range, the range reserved for
Instructional Television Fixed Service (ITFS).   (At California
State University, Chico, the meaning of the acronym is al-
tered slightly, as the initials indicate Instructional Television
For Students.)

    Originally, the ITFS system provided two-way com-
munication between California State University, Chico and the
University of California, Davis, a distance of 92 miles.   Since
1975 the system has expanded.   An integrated network of 13
additional remote sites, two sites in the planning stages, and
two sites scheduled for implementation during 1982-83, now
comprise the CSU, Chico ITFS network.   Currently, all sites
are one-way video, two-way audio, live and interactive.   Four
channels are available for broadcast use; however, only one
television broadcast studio has been developed.   This studio
comfortably seats 32 students.   Four television cameras are
used.   Two pick up the instructor, one provides an overhead
view of the instructor's desk (for example, visuals, and demonstrations), and one televises the on-campus students to the
remote site classes.   One wireless microphone is used by the

instructor; 16 classroom microphones are available for student interaction with off-campus students.

In spring 1975 there were 22 off-campus enrollments for eight course offerings. By spring 1982 there were approximately 400 off-campus enrollments for 25 course offerings. Originally, in 1975, only self-support courses were offered via ITFS, and students paid extension fees to receive credit. In the spring of 1980, a decision was made to provide regular state-supported courses via ITFS. Off-campus students were able to register and pay the same fees as students on campus.

In the beginning, course offerings were in a variety of areas. As the program expanded, the major curriculum components for a Bachelor of Arts in Social Science or in paralegal learning have been made available via ITFS. Plans are under way to provide a minor in Psychology; Business and Electrical Engineering degrees and Computer Science and Education courses are also available.

Five levels of off-campus learning centers have developed. Each level houses varying amounts of electronic equipment. There are 13 Regional Learning Centers. The closest one is at Red Bluff, 42 miles from Chico; the most distant RLC is in Yreka, 173 miles from Chico. An RLC is usually located on a community college campus and has a full complement of equipment, facilities, and direct services staffed on a full-time basis.

The Community Learning Center is usually located in a community public library. It has a smaller complement of equipment and services, and may be staffed at less than full time. The Neighborhood Learning Center may be found in a public school, probably staffed by volunteers. A Home Learning Center has less equipment, facilities, and services. Connection with the television classroom is through a "Brown Box," the connecting device to switch a personal television set to the mid-band ITFS classroom channel, rented from local cable companies. Audio return is by telephone. This Home Learning Center does have the personal advantages of no transportation time or costs and the privacy of the student's home. The Special Learning Centers are located to serve the special needs of the intended students. Equipment and facilities will be planned to provide the special educational needs of the audience.

All CSU, Chico ITFS classes are broadcast live.
Classes are videotaped during broadcast to be available in
the event of technical failure.   If a failure occurs anywhere
in the system, videotapes are shipped to the site.   The students view the taped materials, and the tape is then returned
to Chico and erased.   No attempt to compile a collection of
videotaped classes for rebroadcast has been developed.

Instructional Television For Students at Chico is a
university endeavor; courses from every department are
offered via the system.   The ITFS/Microwave system is co-
ordinated by the Center for Regional and Continuing Education.
Registration, test proctoring, delivery of course materials
to students, library services, and advice for the off-campus
students are all handled by the Center.   Every effort is made
to help off-campus students feel as if they were attending
live classes through the ITFS network.   CSU, Chico, through
technology, welcomes residents of northern California into
the extended classroom program for enrichment, educational
advancement, and enjoyment.

REFERENCES

California Commission on Extended Education.   Telecommuni-
    cations Applications for Extended Education in the Cali-
    fornia State University and Colleges, Long Beach, CA
    1980.   Contact: The California State University, Office
    of Extended Education, 400 Golden Shore, Long Beach,
    CA   90802.

Hall, W. W.   Northeastern California Higher Education Study:
    A Report Prepared for the California Rural Consortium
    and the Coordinating Council for Higher Education, 1972.
    Contact: The Public Telecommunications Project (Calif-
    ornia Public Broadcasting Commission), Center for Com-
    munications, San Diego State University, San Diego, CA
    92182.

ITFS Student Handbook, n. d.   California State University,
    Chico.   Contact the Office of Regional and Continuing Edu-
    cation, California State University, Chico Campus, Chico,
    CA   95929-0250.

Meuter, Ralph F. ; Leslie J. Wright; and Charles F. Urbano-
    witz.   "Closed Circuit Educational Television (ITFS) in
    Northeastern California: The 33,000 Square Mile Cam-
    pus. "   Center for Regional and Continuing Education,

California State University, Chico, California. A Paper prepared for the World Future Society's Fourth General Assembly, "Communications and the Future," Washington, D. C. July 18-22, 1982. Contact: World Future Society, 4916 St. Elmo Avenue, Bethesda, MD 20814.

Myers, A. "Statistical History of ITFS," E & ITV (Educational and Industrial Television), 9(11), (November 1977): 67-70. C. S. Tepfer Publishing Company, Inc., 51 Sugar Hollow Road, Danbury, CT 06810.

Quimby, J. P., and others. Telecommunications and the Public Interest: Needs and Prospects in California, 1974. California Legislature, Report of the Joint Committee of Telecommunications, Sacramento, California. Contact: Public Telecommunications Project (California Public Broadcasting Commission), Center for Communications, San Diego State University, San Diego, CA 92182.

Urbanowicz, C. F. "University Television in Northeastern California: A Partial Solution for the Future?" Paper presented at the "Appropriate Technologies" Session of the Meeting of the Education Section of the World Future Society, University of Houston, Clear Lake City, Texas, October 20-22, 1978. Contact: World Future Society, 4916 St. Elmo Avenue, Bethesda, MD 20814.

Urbanowicz, C. F. "The Department of Education at California State University, Chico, and ITFS (Instructional Television For Students) Involvement: Spring 1975-Spring 1981." Presented at the Meeting of the Program Committee of The Commission for Teacher Preparation and Licensing, Sacramento, California, February 5, 1981. Contact: The Office of Regional and Continuing Education, California State University, Chico Campus, Chico, CA 95929-0250.

Urbanowicz, C. F., and L. J. Wright. "Diversity in Northeastern California: Television as a Partial Solution to the Solution." For the Meeting entitled, "The Next Millennium: Unlearning the 20th Century" for the Education Section of the World Future Society, University of Massachusetts, Amherst, Massachusetts, November 6-8, 1980. Contact: World Future Society, 4916 St. Elmo Avenue, Bethesda, MD 20814.

XV. CALIFORNIA STATE COLLEGE, STANISLAUS

☐   In the fall of 1981, California State College, Stanislaus began broadcast of an interactive television system with one-way video, two-way audio. Transmission is by ITFS with signals covering 10,000 square miles of central California not served by any other public four-year college or university. ITFS will eventually connect the studio classroom at CSC, Stanislaus with four community colleges, twelve remote learning centers, and individual homes in six counties: Stanislaus, Tuolumne, Calaveras, San Joaquin, Mariposa, and Merced.

The geographical terrain of these counties inhibits rapid travel from one location to another. Attending an on-campus class at CSC, Stanislaus involves hours of time in transportation, plus much expense. The opportunity to study via interactive television is very appealing to students and businesses, since employees may work full time and also learn additional business skills without leaving the community.

Eight courses were offered in the fall of 1981, fifteen courses in the spring of 1982, and eighteen courses were scheduled for fall of 1982. Courses are targeted for students interested in returning to school for bachelor's and master's degree programs. The institutions involved include sister campuses in the California State University and Colleges System, and a maximum-security state prison, Deuel Vocational Institution at Tracy. Hospitals and businesses are also interested in acquiring equipment for receiving television classes via the ITFS network.

The broadcast studio at CSC, Stanislaus is housed in a converted physical education building. One studio provides television for the system, utilizing three video cameras. A Faculty Sub-Committee on Instructional Television has recommended to the administration an expansion of the interactive television system to include four studio classrooms and eight additional off-campus learning centers by 1984. Plans for the renovation of the remaining area of the converted physical education building are under way. This facility will be able to house two studio classrooms, engineering requirements, and necessary office space. Long-range plans include the recruiting and training of learning center coordinators at the receiving centers.

Strong administrative leadership has aided the growth of the ITFS program at CSC, Stanislaus. Affirmative faculty interest has been encouraged through communication with faculty and utilization of the distribution and return of needed instructional materials through the highway 99-49 library van delivery system. Every effort is made to continue and support the high quality of teaching standards the faculty has maintained during the 20-year development of the college. Faculty input has contributed to the expansion of the ITFS network, and faculty representatives are involved with ongoing activities as the program expands.

Funding for the interactive ITFS system has developed from various sources, and regional learning centers are being funded through local efforts. Some monies were available from Continuing Education and the Instructional Support Budget at CSC, Stanislaus. Student Services funds were provided for connecting services. The Wells Fargo Bank contributed over $17,000 for equipment at the Sonora Television Learning Center. Viacom Cablevision provided $10,000 to receive a signal at the head-end in Sonora and run a coaxial cable from the top of the mountain to the television learning center two miles away in the Tuolumne County Office Building. The State Department of Corrections funded the Deuel Vocational Institution Television Learning Center at a cost of $10,000. A Continuing Education Grant from the Chancellor's Office of the California State University and Colleges System was obtained to establish the system. In early 1982, a United States Commerce Grant of $83,000 was awarded to expand the mountain network. The ITFS program at CSC, Stanislaus is included in the California statewide "backbone" interconnect program, which is now in the process of development. Within a year this "backbone" system will provide interactive television from Bakersfield to Chico with many CSU and CSC campuses in between on line (Chico, Sacramento to Stanislaus, Fresno, and Bakersfield). Continued expansion is planned through 1985. Positive interest by administrators, faculty, students, and funding organizations indicates that the CSC, Stanislaus ITFS network will continue to enlarge and provide quality learning experiences for many, many citizens in California through the 1980's and beyond.

REFERENCES

Coleman, Laura S. "CSCS's TV Station on the Air," The
    Signal: The Newspaper of the Students of California
    State College, Stanislaus, 22(3), Tuesday, September 29,
    1981. Instructional Media Center, California State Col-
    lege, Stanislaus Campus, Turlock, CA 95380.

Flynn, Helen. "Microwaves Put Students in Class," Stockton
    [Calif.] Record, Monday, August 24, 1981. 530 East
    Market, Stockton, CA 95202.

Goldsmith, Steven. "'Electronic Highway'--Stanislaus State
    Comes to Sonora," The Union Democrat (Sonora, Califor-
    nia), Friday, November 27, 1981.   84 South Washington
    Street, Sonora, CA 95370.

Goldsmith, Steven. "$3,000 Away: 'State College Without
    the 100-Mile Commute,'" The Union Democrat (Sonora,
    California), Wednesday, September 23, 1981.   84 South
    Washington Street, Sonora, CA 95370.

"Grant to Finance Sonora Classes Via Microwaves, "Turlock
    Daily Journal (Turlock, California), Wednesday, August
    12, 1981. P. O. Box 800, Turlock, CA 95380.

"Our own major college," The Modesto Bee, 104(319), Sun-
    day, November 15, 1981. The Modesto Bee, 14th and
    H, Modesto, CA 95352.

Interviews with Dr. Mel Nickerson, Coordinator, Instructional
    Media Center, California State College, Stanislaus Cam-
    pus, Turlock, CA 95380.

☐   In December 1973 the Irvine Unified School District,
Irvine, California, began development of a two-way cable
television system linking 24 school buildings in the district,
the public library, the city hall, the local art museum, and
the University of California, Irvine.   The school district owns
the television system, and the cable company provides transportation of the communication link.   The educational component has been developed using a decentralized system.   Two
dedicated video channels carry audio and video signals to the
schools; channels are provided free of charge.   Each school
is capable of transmitting its own video and audio signal over
one of these leased channels to any school or combination of
schools in the district.   If desired, a third access channel
can be used that allows any CATV subscriber in the community viewing privileges and the ability to interact via conventional telephone lines.

The development of the decentralized system has provided for greater flexibility and use by students.   All equipment is inexpensive and consumer quality.   Students are in
control of the system, including the organizing and planning
of programs, and the setting up and operating of cameras,
microphones and videotapes, channel selectors, and signal
modulators.   The technical apparatus of the two-way cable
system is designed to be understood and operated by an
eight-year-old.   The equipment is extremely durable, and
maintenance problems are minimal.   Students enjoy working
with production equipment and take care of the hardware
rather than abuse it.

Through the first five years of the program many
developmental programs were initiated.   Teachers have used
the system to exchange ideas, offer mini-courses on subjects
of personal interest, conduct in-service training, provide
team teaching in basic skills, and practice role-playing in
the Developing Values program.   Administrators conduct
management meetings via the television system (saving travel
time and costs), televise local experts in learning procedures,

and provide programming to the community to inform citizens
of special school programs. Achievement test results are
sent via cable to the community; only district scores are
discussed. Questions by telephone are answered. School
psychologists conduct guidance clinics via the community
access channel and answer questions by telephone. Psychologists also hold weekly staff meetings for the coordination of
continuing programs at many schools.

Students use the system to share ideas and projects,
and to conduct special research, hobbies, and learning games.
The interactive system encourages productive learning and
information exchange at all levels. By establishing a decentralized cable system the school district has developed an
interactive video system in which the communication technology
provides the following:

- economical one-to-one interaction among learners
  and teachers (participants);

- accessibility to information resources not limited
  to the physical limits of the school building;

- personal control and responsibility for the learning
  function influenced by choice of time and access
  to comparative information resources; resulting in

- fluid time with a personal, rather than a mass or
  group constraint found when classes are scheduled
  for a traditional, specific, pre-planned moment.

Through this decentralized philosophy, an active participatory
system has developed. The originator and the viewer communicate in real time; each has responsibility for content of the
program/interaction; roles change as mobility between creation
and reception of the program content develops, and the interactive process continues.

In the school year 1978-79, the Irvine Unified School
District joined with Control Data Corporation (CDC) in an
experimental program designed to explore the potential uses of
two-way interactive cable television linked with PLATO, the
centrally based computer system now owned by CDC. This
blending added a new dimension to the interactive television
system. During the experiment, information based in the
PLATO computer could interact with live television programming providing data, simulation techniques, projected results,

and many other variations of learning modules.   The CDC is no longer active.

Costs to the Irvine Unified School District were $1, 600 per building as a one time expense for equipment.   Over time, inflation has raised the cost to $3, 000 per building.   With color, the one-time expense for equipment is approximately $4, 000. Categorical grants for programming are used for personnel costs.   These costs seem small when compared to the technology, experiential learning, and communication that students are able to gain.

Irvine Unified School District is looking toward the future.   Every effort is made to ensure students' familiarity and experience with the technology of the present and that of the coming decade.   This technology, including interactive television, is playing and will continue to play a significant role in the personal and business lives of students in the coming years.

REFERENCES

"Report of a Public Hearing Before House Subcommittee on Science, Research and Technology Considering HR 4326, A Bill to Establish a National Commission to Study the Scientific and Technological Implications of Information Technology in Education. " Tuesday, October 9, 1979, 4:00 p.m.   House of Representatives Subcommittee Chambers, Washington, D. C.   Legislative Information, U.S. House of Representatives, House Annex #2, Room 696, Washington, DC   20515.   (202-225-1772)

Ritter, Craig.   "Two-way Cable TV: Connecting a Community's Educational Resources, " Electronic Learning, 2(2), October 1982:60-63.   Scholastic Inc., 50 West 44th Street, New York, NY   10036.

XVII.   CATHOLIC TELEVISION NETWORK, MENLO PARK

☐   The Educational Television Center (ETV) is a member station of the Catholic Television Network (CTN) for the Archdiocese of San Francisco and the Dioceses of Oakland and San Jose.   ETV transmits ITFS signals via four channels to over 150 viewing sites in the greater San Francisco Bay Area.   Signals project in a radius of 100 miles north to Marin County, east to Sacramento, and south to San Jose. Schools, senior citizen centers, hospitals, and religious institutions are included in this network.   ETV services include teleconference facilities; access to a library collection of more than 3,000 videotapes; a production studio for agency, group, or individual program development; and a daily video news show for diocesan information.   Transmission provides one-way video and two-way audio.

The ETV Center began in 1968, and since that time has serviced 135 sites in the San Francisco Bay area.   Service to these schools provides videotaped programming in traditional one-way transmission.   Recent improvements in equipment (such as a transmitter repeater on Mt. Diablo), new playback recorders, reorganization in studio operations, and new studio facilities have supported this service to the schools. In addition, Pastoral Communications services continue to develop programs for parish utilization, including increased evening programming.

The ETV Center has been a pioneer in developing new legal and engineering mechanisms in the communications field.   As an outgrowth of this pioneering effort, the ETV Center at Menlo Park became the first ITFS system to do the following:

- Interconnect major-market CATV systems to create a regional educational CATV network;

- Implement signal-scrambling techniques to control access to sensitive programming in health-related areas;

- Field-test a low-cost satellite receiver for the Japanese National Broadcasting Network (NHK), which was a forerunner of direct home satellite receivers;

- Promote the development of low-cost ITFS receivers in cooperation with manufacturers;

- Interface ITFS with satellites for low-cost ground distribution.

The ETV Center has become involved in many projects. This involvement supports a primary goal of the center to develop technical expansion targeted for diverse audiences. Major projects through the years have included these:

1) PROJECT INTERCHANGE, which utilized the NASA/Canadian Technology Satellite ("CTS"), with ground distribution through ITFS and CATV connections. Teachers in the Archdiocese of San Francisco and Torrance Unified School District (Los Angeles) teleconferenced with the resource staff of California State University at Chico and with the office of the Santa Clara County Superintendent.

2) Other teleconferences focused on special education. Through the National Institute of Education, the ETV Center worked with the Council for Exceptional Children of Reston, Virginia, and with the University of Kentucky to develop national satellite-based service in special education.

3) The SENIOR UNIVERSITY was developed through a grant from the Fund for the Improvement of Post-Secondary Education. SENIOR UNIVERSITY is a self-supporting model network that provides education and information to the senior population of the San Francisco Bay Area.

4) The HEALTH CARE NETWORK sent continuing educational programming to staff members in fourteen area hospitals. Program sources included the Veterans Administration, the National Library of Medicine, and locally produced or leased programming.

5) The SPECIALIZED SCIENCE NETWORKS was

developed through a contract with NASA/Ames.
The ETV Center interconnected over 300,000
homes in northern California with a special CATV
system. Specialized programming originated
from museums and universities, as well as through
transmission of "live" programs from the Pioneer Missions to Venus and Saturn.

6) Current special projects are concerned with family communications programming and with developing television services for the public service
community.

In addition to the special projects, many satellite demonstrations have been conducted through the ETV Center, covering
a wide range of subjects including local government issues,
instructional reading, and artistic applications.

As a result of the on-going search for new applications
of technology, many non-school clients have used teleconferencing services, both regionally in the San Francisco Bay
Area and nationally through satellite/ITFS interface. The
experience has demonstrated that teleconferencing is most
effective when the following conditions prevail:

- the primary purpose of the conference is for the
  exchange of information (negotiation or conflict
  resolution should not be the primary goal);

- conferences are held on an on-going basis to exchange information;

- the video portion of the conference provides visuals
  which enhance the information exchange;

- participants have met face-to-face at some time
  prior to the teleconference;

- the teleconference acts as a follow-up meeting to
  previous contacts.

The Educational Television Center of the Catholic Television Network has pioneered in the telecommunications industry through developing a policy to utilize new technologies and
develop new applications for specific users. This policy has
allowed many segments of society -- government, business,
social, educational, and religious -- to benefit from these pilot

programs. In time, all of society will gain from the new communication formats developed in the Menlo Park area and shared throughout the world.

REFERENCES

Catholic Television Network. (Descriptive Flyer, no date.) Educational Television Center, 324 Middlefield Road, Menlo Park, CA 94025. (415-326-7850)

Elementary Broadcast Summary, Educational Television Center, Bay Area Dioceses, September 1981. Catholic Television Network, San Francisco. Educational Television Center, 324 Middlefield Road, Menlo Park, CA 94025. (415-326-7850)

Green, David, and Bill Lazarus. "A User-Controlled Teleconference Studio," E & I TV (Educational and Industrial Television), 12(4), April 1980. C. S. Tepfer Publishing Company, Inc., 51 Sugar Hollow Road, Danbury, CT 06810. (203-743-2120)

Special Projects of the Educational Television Center, ETV Center, no date. Educational Television Center, 324 Middlefield Road, Menlo Park, CA 94025. (415-326-7850)

☐   Brown University, Providence, Rhode Island, has developed an extensive broadband communication network that
is providing reliable communication for data transmission
and television throughout the campus.   More than 110 buildings are connected currently; in time, all buildings on campus
will have electronic interconnects.

Developments of this communication system, called
BRUNET (BRown University NETwork), began in June 1980
with the appointment of a Study Group on Telecommunications
and Networks. Within five months, a determination of the primary requirements included the need to connect the many new
terminals on campus to the time-sharing computers more
efficiently, and the need to provide reliable communication
for the new energy management and security applications.

A broadband communications cable network was selected to meet the major purposes of 1) providing a "long-haul"
system for connecting different local networks of personal computers and intelligent stations in various departments of the
university; 2) constructing a utility providing the entire university, both on- and off-campus, with ready access to shared
computing, information, and communication resources; and
3) developing a framework for orderly growth from a central-
ized computing environment to one in which resources are dis-
tributed.   The first major task was the construction of an
electronic "backbone" to serve as the main communication
skeleton.   The network "backbone" consists of two 300-MHz
mid-split CATV systems installed completely underground.
From this "backbone" (primary distribution system) the
branch connections can be installed gradually. Estimates
indicated that the primary distribution system could include
31, 000 feet of dual trunk cables (62, 000 feet total).   In each
of the two redundant systems, 17 trunk amplifiers are used;
the maximum number of amplifiers in cascade is five.   An
additional factor suggesting a gradual development of the
system was the need to examine the true condition of the
existing underground conduits prior to making changes in the
BRUNET cable lines.

The installation of data communication services was well underway when equipment specifications were altered to include additional immediate communication needs. Tariff increases granted to New England Telephone affected Centrax users. Costs for Brown University were to increase by 45 percent. This increase in maintenance costs triggered a rapid transfer of the telephone system, switching to a Bell Dimension 2000PBX system. Unfortunately, not all of the data traffic was considered in the design, and additional wiring for voice (telephone) users was required. A new date for changeover to the Dimension telephone system was established and met in February 1982.

The communication system was designed by dividing the campus into four quadrants. The head end is situated in a central location with respect to the conduit system housing electrical connections. An independent and redundant trunk line serves each quadrant, with the trunk lines from each redundant system combined in the head end for certain applications. Campus security offices are located near the head end, facilitating electronic security television reception and the monitoring and alarm systems.

In addition to the security systems, there is an extensive energy management system. Terminal-to-host communication and host-to-host computer systems have been installed. An off-campus gateway to access computers between the network and the telephone system was established. An on-campus telephone system is operating efficiently at a $20 per port monthly rate--one way of making the on-campus telephone system cost-efficient. Over 750 terminals, computer ports, and other internally wired devices were connected during the first phase of network development. All equipment components and systems selected for installation are commercially available for ease of maintenance, installation, and cost.

The State of Rhode Island has developed extensive plans and requirements stating that two-way communication will go to all public and private institutions. Requirements also include an interconnection among all cable franchises, of which there are nine in Rhode Island. BRUNET is aware of these requirements, and all equipment is installed to specifications that meet these long-range developments.

The technology for interactive television classes on-campus is in place. However, Brown University first plans to implement interactive instructional television at 20 hospitals located throughout Rhode Island. Conserving transportation time and costs is important to the medical personnel in the state. Brown University, through the BRUNET system, will be able to meet this need.

REFERENCES

Shipp, William S., and Harold H. Webber. "Wiring a University," Institute of Electrical and Electronic Engineers, International Conference on Communications (ICC), June 13-17, 1982. Contact Institute of Electrical and Electronic Engineers, Inc., 345 East 47th Street, New York, NY 10017.

Interview with Dr. William Shipp, Brown University, Providence, RI 02912.

GLOSSARY

BIBLIOGRAPHY

APPENDIXES

INDEX

GLOSSARY OF TERMS

ANALOG: Analog transmission is required for sending the
audio and video signals of television. With analog transmission the video picture is represented through the ma-
nipulation of an electronic signal. The quality of the television picture is very dependent on the quality of the
electronic signal.

AUDIO: The sound track sent via a telecommunications
system. Many voice-grade channels (approximately 40-
400 Hz each) are possible. Voice-grade channels can be
used for telephone trunking, conferencing between campuses, talkback on microwave television courses, facsimile, or other special narrow-band communication,
depending upon the specific equipment installed.

AUDIOVISUAL INSTRUCTION: An almost historical term
indicating the use of materials that employ visual and/or
auditory stimuli to aid in the learning process; these
materials may include films, television, remote site
auditory response systems, as well as other visual and
auditory mechanisms.

CABLE TELEVISION: A system of sending television signals
through coaxial cable lines. This system of transmission developed in the late 1940's. Cable television transmission is necessary where many physical obstacles
(mountains, valleys, high buildings) impede the clear
transmission of reception signals. Cable transmission
is also used to provide a greater selection ("menu") of
television programming. Cable television has two primary means of transmission:

Community Antenna Television (CATV): Transmission
is sent over cable television to all subscribers; reception is not limited to schools, or to selected channels.

Reception is available to all people in the cable subscription area, whether interested or concerned with
learning via television.   No "privacy" in teaching can
be assured.

Closed-circuit Television (CCTV):   Transmission is by
way of coaxial cables connected only to the school
buildings and/or institutions scheduled to receive instruction via television, or through channels reserved
for use only by the institution.   Closed-circuit may
be between rooms within one building, and/or between
buildings (schools) for a large school district, or between school districts, if the buildings have been
equipped with the proper receiving equipment.

Off-campus or city-franchised cable television systems
can effectively extend educational outreach at low cost.
Homes and schools can be served by one-way television;
talkback can be provided by several methods.   Each system has individual requirements that are met with specific applications of technology.

On-campus television includes inter- and/or intra-build-
ing cable distribution systems.   Technology implementation would depend upon the specific system requirements.

DATA: "Data" is a broad term, indicating many kinds of
information sent by digital transmission.   Definitions of
terms used within the Data Communication environment
include these:

Band: The frequencies which are within two definite
limits, and which are used for different purposes.

Bit: A contraction of "binary digit," which is a unit
of information in two-level (binary) notation.   "Bit"
indicates a single character in a binary number.

Bit Rate: The speed at which digital data are transmitted, expressed in bits per second.   Kbit: one
thousand bits per second, called a Kilobit.   Mbit: one
million bits per second, called a Megabit.

Bit Error Rate: The number of errors in transmission due to all causes.   A high number of bit errors
in a system reduces the efficiency of the system.

Channel: A path along which information, particularly a series of digits or characters, may flow.

Facsimile: A printed paper copy of teletext information, also called "hard copy." The transmission process for facsimile commonly uses rotating drums at the transmitter and receiver. The transmitter scans printed material, photographs, or drawings, and converts a varying electrical signal to voice frequencies for transmission. The receiver reconverts the voice frequencies to signals that drive a stylus for recording the facsimile copy, or uses the same signal in a photographic recording process.

Modem: A contraction for the term "Modulator-demodulator." This is a terminal device with common and cost-saving circuit elements. It allows digital data to be transmitted over voice-oriented (analog) communication facilities. A modem is used to connect individual terminals to a main frame computer, or to a computer data bank.

Multiplexer: A device that allows for the transmission of a number of different messages simultaneously over a single circuit. Usually the incoming messages are at relatively low transfer rates and the single outgoing circuit is at a high transfer rate or speed.

Node: The point of a network at which many links terminate. It may or may not be a switching node.

Queue: A series of potential users waiting for outgoing transmission lines and/or processing. This is sometimes referred to as the "back-up."

Synchronous: In data communications, this is a system whereby the transmitting and receiving terminal equipment is kept in step by timing signals, whether or not the path is in use.

Television: A signal system which converts transient (lively) visual images into an electrical waveform for transmission over Radio Frequencies via satellite, microwave, ITFS, or cable to receivers where the waveform is converted back to the original image. The visual image is accompanied by an aural signal (voice, music, or other sound).

Slow-scan television is a special form of visual
transmission.   Images are not lively, and complete
pictures are transmitted from a few seconds to thirty
seconds apart.   The advantage is that the frequency
spectrum transmission space is reduced.   This system could be described as a "live slide projector."

Television sources for sending programs can be
live, as with classrooms, conferences, or seminars;
or programs can be prerecorded using videotape, film,
slides, video-graphs, or charts.

Other media: Electronic mail, library automation,
and other special electronic transmission methods are
all specific kinds of data providing forms of interactive communication.   The only requirement for using these forms of data is the installation of the proper
terminal equipment (transmitters, receivers, printers,
etc.).

Timesharing: The use of a device (i.e., a computer)
for two or more purposes during the same overall time
interval; this is accomplished by interspersing component actions in time.

DIGITAL: Digital circuits are needed for hard-copy transmission and for accurate access and retrieval from re-
motely located computer and memory banks.   In the
digital mode the video picture is represented numeric-
ally in much the same way as a computer manipulates
data.   The incoming numerical data are reinterpreted
by the receiver, which then composes the pictures based
on these "data."

DIRECT BROADCAST SATELLITES (DBS): A satellite communications system in which the satellite-to-earth signal (downlink) is powerful enough to allow reception
through an inexpensive, consumer-maintained receiver
(often called a "home dish").

DISH: A ground-based receiving or transmitting antenna,
sometimes shaped like a giant saucer.

DOWNLINK: A satellite antenna system used to receive program signals from a specific satellite; the system is ground-based.

FIBER OPTICS: Transmission of electronic information, including audio/video materials, through light signals passed through extremely small glass fibers.

HERTZ (Hz): A unit of frequency equal to one second per cycle. This unit was named for H. R. Hertz in 1967, and replaced an older, more descriptive term of "cycles per second."

INSTRUCTIONAL TELEVISION (ITV): Any television program developed specifically for instructional purposes; such programs are frequently developed in conjunction with a specific course or set of lessons.

INSTRUCTIONAL TELEVISION FIXED SERVICE (ITFS): In 1963 the Federal Communications Commission (FCC) set aside a part of the microwave spectrum to be used by educational institutions for instructional purposes. This spectrum area, the 2500-2690 megahertz (MHz) band, contained 31 six-megahertz channels. In 1971 the FCC reduced the number of six-megahertz channels from 31 to 28, and further ordered that these 28 channels be used exclusively for non-profit educational purposes. ITFS transmission differs from Operational Fixed television bands in that Operational Fixed Bands may be multipurpose, with two-way television, data, voice, facsimile, and point-to-point. The primary features of ITFS include these:

- ITFS can be broadcast omnidirectionally.

- Institutions are allowed to transmit up to four simultaneous television programs in a given service area, which extends to about 30 miles from the transmitter.

- ITFS response is permitted for two-way audio between remote students and on-campus classrooms. Response can be voice (talkback) or data.

- ITFS channels retain a United States standard broadcast format.   Simple microwave down converters
  at the remote classroom deliver VHR signals to
  ordinary TV receivers.

- Special permission from the FCC can extend ITFS
  capability (higher power transmitters for greater
  range, special sub-carrier frequencies for other
  media, telemetry, etc.).   However, in the basic
  design, ITFS provides one-way video and two-way
  audio in a broadcast area.

- Signal path considerations follow the laws of physics, similar to those described in the microwave
  section (see below).   Frequency assignments and
  technical matters are governed by Part 74 of the
  FCC Rules and Regulations.

- ITFS capability can assist in teaching students in
  one or more geographically remote classrooms on
  the basis of direct face-to-face relationships.

- ITFS can deliver "written" materials to remote
  locations from a central point.

- ITFS has low-cost digital circuits available to
  access, control, and distribute remote computa-
  tional power and data-bank information.

LOW-POWER TELEVISION: Television signals transmitted
on any channel between 2 and 69; maximum power is 100
watts (VHF) to 1,000 watts (UHF).   The signal range is
12 to 15 miles.

"Originating" low-power television station (LPTV):   A
television station which originates programming without restrictions for all or part of the broadcast schedule.

"Non-originating" low-power television station (often
called a TV translator):   A television station which
rebroadcasts the signal of a full-service television
station.   However, one break of no more than 30 seconds per hour may occur.   During this 30-second
break, the station may solicit or acknowledge financial support.

MICROWAVE: Radio transmission at frequencies of 1,000 MHz or above.   The physics of transmission requires that no intervening objects (such as mountains, buildings, trees) obstruct the line-of-sight path.   For maximum path reliability and reasonable cost, path clearance over objects of at least 60 percent of the First Fresnel Zone is required.   This zone can be described as a symmetrical volume around the line-of-sight path (path of the direct wave).   Any point inside the zone can be considered an imaginary reflection point, but the delay of the reflected wave, relative to the direct wave, shall not exceed one-half wave length.   In addition to 0.6 of the First Fresnel Zone, transmission designers should also be concerned with earth curvature, refraction, reflection, attenuation, and other atmospheric phenomena.

Digital microwave: All modulation on the microwave carrier frequency is digital in format.
Audio -- The human voice must first be "digitized," in other words, the varying (analog) voice levels must be changed to a coded series of digits for transmission; then, at the receiver, the series of digits must be decoded to varying voltage levels in order to drive a speaker or earphones.
Data -- In general, all modern computers are digital in format, and data may be transmitted without conversion.
Video -- Standard visual images produced by a camera (or recorded for later transmission) are, like the human voice, analog in format.   That is, the camera or playback recorder produces a varying level of voltage.   To transmit this complex analog signal over a digital microwave system requires coding (in other words, the picture must be "digitized").   At the receiver, the digital signal is decoded to a varying voltage to drive a picture monitor.   Similar to the MODEM (modulator-demodulator) the CODEC (Coder-decoder) performs the necessary conversion from analog to digital, or vice versa.

Analog microwave: Amplitude and frequency modulation of the microwave carrier frequency can be by a mixture of digital and analog signals.

NARROWCASTING: A television industry term indicating that the content of the program is designed for a very specific

audience, often content that the general public might not
wish to watch. Examples would be programs using a non-
English language for the dialogue, programs concerned with
specific technical equipment or content, or programs using
medical terminology to describe detailed methods of treatment or surgery for rare conditions or diseases.

OTHER TRANSMISSION SYSTEMS: Special new and innovative
transmission systems are being developed. Fiber optics,
for example, holds great promise because of its broad
information handling capability. This is a physical system
for transmission (like cable, it must be placed between two
points of communication), and some time will elapse before
there is wide implementation of fiber optics. Even then,
fiber optics may only be used over dense communication
paths.

PLATO (Programmed Logic for Automatic Teaching Operation):
A large-scale, computer-based system developed at the
University of Illinois. Initially, PLATO provided instruction ranging from the college to elementary level. The system utilizes a plasma-display panel which retains images
and responds directly to the digital signals from the computer, in contrast to the cathode-ray tube display on which
images must be continually regenerated. The PLATO system has been purchased by Control Data Corporation for
further development in the commercial market.

SATELLITE: Satellites are free standing space ships, custom-
designed to travel great distances beyond the atmospheric
circle surrounding Earth. Satellites, depending upon specifications, may be operated by remote control or by on-
board personnel. Satellites are geostationary or synchronous; many are in orbit around Earth. The orbit is circular;
the distance from Earth to the satellite is approximately
22,300 miles. If the satellite is launched to the east (the
direction of Earth's rotation), the satellite will make a
complete orbit around the Earth in 24 hours. It may appear
to stand still over one point on Earth.

Communication through the use of satellites has developed during the past fifteen years. Communication capability has increased from audio return to video transmission
and many forms of data-return and slow-scan television.

Satellites are used by many areas of business, industry, and government.

TELECONFERENCE: A closed video transmission, often via satellite. Participants have the ability to communicate with other participants located at the original point of transmission or at other remote sites of broadcast.

TELETEXT: This is the generic name for a broadcast service that uses several scanning lines of the vertical interval between frames of a television picture in order to transmit alphanumeric information to home television receivers, along with the standard television picture. This system requires that printed material be stored in a computer. When the information is desired for broadcast, the alphanumeric print is transmitted onto the television screen, either with a plain background or superimposed over another television picture. Specific titles for teletext information are used by specific organizations. Examples include CEEFAX, used by the British Broadcasting System, and ORACLE, used by the Independent Broadcasting Authority in Britain.

TRANSPONDER: The receiver-transmitter portion of a communication satellite. The transponder receives signals from the ground, amplifies the signal, and transmits the signal back to Earth.

UPLINK: A ground-based transmission station which sends programming to a satellite.

REFERENCES

Many of the definitions in this Glossary of Terms have been provided through the kind assistance of J. Patrick Loughboro, President of JP Associates, Inc., Torrance, California. Particular reference should be made to Mr. Loughboro's report:

University of California, Berkeley: A University Wide Telecommunications System, January 1982. Contact: JP Associates, 3115 Kashiwa Street, Torrance, CA 90505 (213-539-8533)

Additional information in the area of ITFS was derived from:

Curtis, John A., and Joseph M. Biedenbach. Educational
    Telecommunications Delivery Systems, 1976. The Amer-
    ican Society for Engineering Education. One Dupont Cir-
    cle, Suite 400, Washington, DC 20036.

Many definitions in this Glossary were enhanced through information in:

Educational Technology: A Glossary of Terms, 1977/79.
    Association for Educational Communications and Tech-
    nology. Publications Department, Association for Edu-
    cational Communications and Technology, 1126 Sixteenth
    Street, N.W., Washington, DC 20036.

BIBLIOGRAPHY

☐   A great many experimental programs and isolated uses
of interactive television have developed during the past two
decades. The following bibliography contains sources--not
mentioned earlier in the text--for many of these experi-
mental/pilot programs. Interactive television has many formats useful for a variety of purposes in all sectors of society. As technological developments and economics allow,
these experimental programs will be introduced and utilized
in the mainstream of our society. This bibliography is included as a guide to aid readers in learning more about this
new format of communication.

GENERAL

Almes, Steven. Telecommunications: A Picture of Change,
    An Upper Midwest Report, May 1981. Upper Midwest
    Council, 250 Marquette Avenue, Minneapolis, MN   55480.
    A nontechnical report on the equipment and the uses of
    telecommunications, in late 1980 and early 1981 in Iowa,
    Wisconsin (the upper peninsula), Nebraska, North Dakota,
    South Dakota, Montana, and Minnesota.

Arlen, Gary. "Plugging-in Coral Gables, " American Film,
    January/February 1980. American Film, John F. Ken-
    nedy Center, Washington, DC   20566.
    Viewdata Corporation of America selected 200 fam-
    ilies in Coral Gables, Florida, to participate in an ex-
    periment to determine consumer use of a television in-
    formation retrieval system. Applications of viewdata and
    teletext are included to provide a variety of information,
    including real estate prices, airline schedules, and news
    reports.

Baer, Walter S. Interactive Television Prospects for Two-
    way Services on Cable, November 1971. A report pre-
    pared under a Grant from the John and Mary R. Markle
    Foundation. Rand Report R-888-MF, The Rand

Corporation, 1700 Main Street, Santa Monica, CA 90406.
This report describes the development of two-way interactive communication services on cable television systems.
Specific chapters relate to interests of education, business,
and other sectors of society.

Baldwin, Thomas F.; Thomas A. Muth; and Judith Saxton.
"Public Policy in Two-way Cable: Difficult Issues for a
Developing Technology Telecommunication Policy," June
1979. Butterworth Scientific Limited, Journals Division,
Box 63, Westbury House, Bury Street, Guildford, Surrey
Gu2 5BH, England.
The authors describe the policy of the United States
in 1979 regarding two-way cable communications, sug-
gest policies for implementing service, and outline responsibilities of local franchising authorities specific to
two-way television services.

Baran, Paul. "30 Services That Two-way Television Can
Provide," The Futurist: A Journal of Forecasts, Trends,
and Ideas About the Future, October 1973. World Future
Society, 4916 St. Elmo Avenue, Bethesda, MD 20814.
Description of thirty specific services that interactive
television can provide, based on a 1970 study by the
Institute for the Future.

Baruch, Jordan J. Interactive Television: A Mass Medium
for Individuals, 1969. Corporation for Public Broadcasting, Washington, DC. (ERIC Document Reproduction
Service Number ED 057 609)
Description of System-3, a one-to-one communications
network using receiver-terminals and interconnections of
System-3. Many applications are included, as well as
organizational, regulatory, and technological considerations needed before System-3 can be realized.

Burnham, David. "The Twists in Two-way Cable," Channels
of Communication, June/July 1981. Media Commentary
Council, Inc., 1515 Broadway, New York, NY 10036.
Computer-response television may affect the quantity
of recorded data collected about the personal habits of
the viewers. Multichanneled opportunities for televised
information and entertainment will have an impact upon
subscribers, as well as on those who do not participate
in cable television services.

Carey, James W. "Changing Communications Technology

and the Nature of the Audience" (The 1980 Journal of
Advertising Invited Luncheon Address), March 23, 1980,
Columbia, Missouri, Journal of Advertising, 9(2) (1980):
3-9+. American Academy of Advertising, Anthony F.
McGann, Editor, Box 3275, University of Wyoming, Lar-
amie, WY 82071. (307-766-1121)
    This article provides a review of the changing meth-
ods for public communication. Changing communication
technology, since the advent of television, has altered
the nature of the receiving audience. Many individuals
have expressed concern that society is facing a communication revolution with information assembled and dis-
seminated to larger masses than ever before, resulting
in conformity and endless repetition. However, the
author indicates technology has not caused a revolution,
but rather an opportunity for audience interest and increased participation in specific content which has contributed to the development of narrowcast programming.

Dowlin, Kenneth E. "CATV + NCPL - VRS," Library Jour-
    nal, 95(15), (September 1, 1970):2768-2770. 1180 Avenue
    of the Americas, New York, NY 10036.
    Describes the video reference service created by Com-
    munity Antenna Cable Television through the Natrona
    County (Wyoming) Public Library. A step-by-step pro-
    cedure for implementation of a reference service is de-
    tailed.

Hickey, Neil. "Read Any Good Television Lately?" in Barry
    Cole, ed., Television Today: A Close-up View, 1981.
    Oxford University Press, 200 Madison Ave., New York,
    NY 10016.
    Review of printed alphanumeric systems that may be
    available for cable television subscribers in the United
    States.

"Home Telecommunications: Paths for Growth in the 1980's,"
    The Futurist: A Journal of Forecasts, Trends, and
    Ideas About The Future, June 1980. World Future Soci-
    ety, 4916 St. Elmo Avenue, Bethesda, MD 20814.
    Comments on the growth of home telecommunications
    during the current decade. Some telecommunication use
    is based on the interactive television study conducted in
    Reading, Pennsylvania.

Jenkins, John A. "The Conscience of Cable," TWA Ambas-
    sador, 15(3), (March 1982):51-62. The Webb Company,

1999 Shepard Road, St. Paul, MN   55116.
Examination of the privacy issue involved in cable
subscription.   Franchise holders have few regulations
at present.   As two-way television capabilities prolifer-
ate, lack of regulations may increase the likelihood of
privacy invasion.   Now is the time for citizen action
with regard to franchise regulations.

Jones, Martin V.   "Social Impact of Interactive Television,"
IEEE Transmission Communication, COM-23(10), (October 1975):1156-1163.   Institute of Electrical and Electron-
ics Engineers, Inc., 345 East 47th Street, New York,
NY   10017
This paper forecasts changes that may result when
interactive television is widely accepted in society.   Changes will occur in economic, social, political, institutional,
and legal areas; citizens will alter daily routines of work,
home, and leisure activities.

Journal of Communication, Autumn 1978.   Annenberg School
of Communication, 3620 Walnut Street, Philadelphia, PA
19104.
This issue is devoted to articles about interactive television sites in Pennsylvania, Alaska, South Carolina, and
Illinois.

Katzman, Natan.   "The Impact of Communication Technology:
Promises and Prospects," Journal of Communication,
Autumn 1974.   Annenberg School of Communication, 3620
Walnut Street, Philadelphia, PA   19104
Six propositions concerning the relationship between the
new communication techniques and the distribution of in-
formation in society are noted, and major implications
are discussed.

Land, Phyllis.   "A New Breed for the 80's: Irvine's School
Library Media Program," School Library Journal, August 1980.   1180 Avenue of the Americas, New York, NY
10036.
Report of the award-winning School Library Media
Program in Irvine, California.   Includes information about
the student-controlled, two-way interactive television system operating daily between school buildings.

Low, Ken.   "Intoxicants, Human Development and the 80's,"
School Guidance Worker, 35(3), (January 1980).   University of Toronto, Guidance Centre, 252 Bloor Street West,

Suite 4-299, Toronto, Ontario M5S 2Y3, Canada.
Interactive television systems and other learning technologies will increase self-defeating behaviors by viewers.
A hypnotic preoccupation with high-impact diversion may
result in greater use of destructive dependencies, including intoxicants, in the 1980's.

McGillem, Clara Duane, and W. P. McLauchian. Hermes
Bound, 1978. Purdue University Press, South Campus
Courts, D., West Lafayette, IN 47907.
An examination of the relationships between society
and telecommunications technology is presented in this
book. Political, economic, and social factors and tech-
nical constraints are discussed. Chapter Five is entitled,
"Two-way Telecommunications."

Smith, Ralph Lee. The Wired Nation: Cable TV; The Electronic Communications Highway, 1972. Harper and Row,
10 East 53rd Street, New York, NY 10022.
The information in this text originally appeared as a
magazine article in a special issue of The Nation, May
18, 1970. The text suggests that communication technology emerging during the last third of this century,
under the jurisdiction of the Federal Communications
Commission, will impact upon society. Intelligent deci-
sion-making, leading toward a national communications
system, is examined in this text.

Stetten, Kenneth. "Cable Television for Librarians: Interactive Cable System," Drexel Library Quarterly, 9(1 &
2), (January-April 1973):99-105. Centrum Philadelphia,
University City Science Center, 3624 Market Street,
Philadelphia, PA 19104.
This paper suggests broadening the scope of current
cable television to include a range of services that have
high potential for social impact. Interactive television
can provide the means of communication between subscriber and information source.

Stetten, Kenneth J., and John L. Volk. A Study of the Tech-
nical and Economic Considerations Attendent on the Home
Delivery of Instruction and Other Socially Related Services
Via Interactive Cable TV: Volume I, 1973. National
Science Foundation, Washington, DC 20006 (ERIC Docu-
ment Reproduction Service Number ED 981 219)
The first of four volumes summarizing the results of
the Mitre Corporation's five-year study on interactive

television. Technical, economic, and operational considerations are discussed and future plans described.

Thomassen, Cora E. CATV and Its Implications for Libraries, 1974. University of Illinois Graduate School of Library Science, Urbana-Champaign, IL 61801.
Cable television as utilized by libraries is described. Potential and pilot applications for community service, information referral, and interactive communication are examined.

"TV Turns to Print," Newsweek, July 30, 1979, pp. 73-75. 444 Madison Avenue, New York, NY 10022.
Description of viewdata and teletext systems under consideration for use in the United States.

Vieth, Richard. Talk-Back TV: Two-way Cable Television, 1976. Tab Books, Blue Ridge Summit, PA 17214.
Examines the technology needed to implement many types of two-way television (pay TV, at-home shopping, video games, banking, computer response). Provides information on many early test programs using interactive television.

von Feldt, James R. An Overview of Computers in Education, March, 1977. National Technical Institute for the Deaf, Rochester Institute of Technology, Rochester, NY 14623.
An overview of computers in education, presenting a brief history of computers and highlighting the many uses of computers. Various viewpoints on the use of computers in education are examined.

Wicklein, John. "Wired City, USA: The Charms and Dangers of Two-way TV," The Atlantic Monthly, February 1979, pp. 35-42. 8 Arlington Street, Boston, MA 02116.
Description of QUBE, the two-way interactive cable television system installed in Columbus, Ohio, by the Warner Cable Corporation. Long-range implications of QUBE are discussed.

The Wired Public Library: Who Needs It? What Will It Cost? Who Will Pay For It? 1973. New Jersey State Library, Trenton, NJ 08608. (ERIC Document Reproduction Service Number ED 092 138)
A description of the cooperative study sponsored by Atlantic City and Cape May County that investigated the use of cable and interactive television to extend library

services. In 1973 public interest was high; however, national leaders felt that implementation of the required technology had not yet reached cost efficiency.

Wood, Fred B., and others. Videoconferencing Via Satellite: Opening Congress to the People, Summary Report, February 1978. George Washington University, Washington, DC. Program of Policy Studies in Science and Technology. (ERIC Document Reproduction Service Number ED 162 647)
    Comments from several United States senators and representatives who participated in the videoconferencing project. Personal evaluations plus answers to seven prepared questions. Diagrams, graphs, and photographs included.

Wright, Gwendolyn. "The Future of Video Technology," Educational Technology, 21(4), (April 1981):40. Educational Technology Publications, 140 Sylvan Avenue, Englewood Cliffs, NJ 07632.
    Changes in technology during the next ten years will include "holomorphs"--small plastic chips containing the "complete form" of a video program. Very small television sets plus teletext and viewdata will increase access to information services for everyone.

Zenor, Stanley D. "Turn Your Television into an Information Terminal," Instructional Innovator, 26(2), (February 1981). Association for Educational Communication and Technology, 1126 Sixteenth Street, N.W., Washington, DC 20036.
    Examination of teletext and viewdata applications. Major questions of implementation, plus questions facing industry and commercial networks, are discussed.

BUSINESS

Barilovich, O. I.; A. D. Matveyev; and B. S. Yalkunin. "Design of Simplex Branches of Radio-Relay Links Telecommunication," Radio Engineering and Electronic Physics, 31-32(2), (February 1977):12-17. Scripto Publishing Company, 7961 Eastern Avenue, Silver Springs, MD 20910.
    Branch circuits at microwave frequencies can divert part of the microwave energy from the wave channel of a radio relay trunk line. An additional antenna system will transmit energy in the desired direction. This

design is often used in supplying television programs to
low-power television relay stations.

Barnes, Richard E., and Thomas Reagon. "The Future of
Electronic Entertainment," The Futurist: A Journal of
Forecasts, Trends, and Ideas About The Future, 10(1),
(February 1976):40-43. World Future Society, 4916 St.
Elmo Avenue, Bethesda, MD 20814.
Description of many emerging forms of electronic
communication. The danger of new communication systems may lie in the use of technology to invade privacy
and in the isolation of individuals who might replace
human interaction with electronic stimuli and surrogate
conversationalists.

Bolton, W. Theodore. "A Lesson in Interactive Television
Programming: The Home Book Club on QUBE," Journal
of Library Automation, 14(2), (June 1981):103-108. Journal of Library Automation has been retitled to Information
Technology and Libraries; both titles published by American Library Association, 50 East Huron Street, Chicago,
IL 60611.
Description of the Columbus, Ohio, two-way television
system, specifically describing a book discussion program
aired over QUBE. Initial programs did not generate an
enthusiastic response from viewers; however, future potential interest should not be overlooked.

Brown, Arnold. "Communications, Computer Changes Are
Only Part of the Future Challenge Facing Marketers,"
The Marketing News, 14(12), (December 12, 1980):28.
American Marketing Association, 222 South Riverside
Plaza, Chicago, IL 60606.
Suggestions for changes to adapt to the electronic
technology that is causing vast changes in business com-
munication and recording methods.

Buckelew, Donald P., and David W. Penniman. "The Outlook
for Interactive Television," Datamation, 20(8), (August
1974):54-58. Technical Publishing Company, 875 Third
Avenue, New York, NY 10022.
Description of experimental interactive television programs prior to 1975. Invasion of privacy, potential
applications, and needed research are discussed.

Bushman, I. Anthony, and Richard Robinson. "Two-way
Television: A Tool for New Product Research," Business

Horizons, 24(4), (July/August 1981):69-75. Graduate
School of Business, Indiana University, Bloomington,
IN 47405.
Two-way interactive television has great potential as
a market forecasting tool. The financial cost is similar
to that of telephone surveys or personal interviews. Limitations of interactive television in market forecasting is
also described.

Butler, David W. "5 Caveats for Videodiscs in Training,"
Instructional Innovator, 26(2), (February 1981):16+. Association for Educational Communication and Technology,
1126 Sixteenth Street, N.W., Washington, DC 20036.
Current obstacles to the implementation of videodiscs
are cited. Factors of economics, subject content, trained
staff, and interaction capabilities are mentioned. The
potential value of videodiscs is more significant than the
current obstacles.

Educational and Industrial Television, 13(6), (June 1981). C.
S. Tepfer Publishing Company Inc., 51 Sugar Hollow
Road, Danbury CT 06810.
Special issue devoted to interactive video.

Eller, Timothy S. Mitre Interactive Television Experiment,
Official Transcript of the 23rd Annual National Cable
Television Association Conference, Stockton, California,
1974. National Cable Television Association, 1724 Mas-
sachusetts Avenue, N.W., Washington, DC 20036.
Demonstration of interactive television in which the
Educational Testing Service and Big Valley Cablevision
participated.

An Experiment in Conference TV, 1974. British Columbia
Telephone Company, Vancouver, British Columbia, Canada. (ERIC Document Reproduction Service Number ED
104 425)
Description of an early two-way television conference
conducted between Vancouver and Victoria, British Columbia. Telephone officials, businessmen, government
officials, and college students participated.

"Experiments in Interactive Cable TV," Journal of Communication, Autumn 1978. Annenberg School of Communication, 3620 Walnut Street, Philadelphia, PA 19140.
This issue presents a series of articles about several interactive television projects conducted in the United
States.

Feldman, Nathaniel E. Interconnecting Cable Television Sys-
    tems by Satellites: An Introduction to the Issues, 1973.
    National Telecommunications Information Service, 608
    13th Street, N.W., Washington, DC 20004.
       A review of recent television technologies, including
    two-way television and high-power satellites.

Friend, David. "Color Graphics Information Systems Boost
    Productivity," Mini-micro Systems, 13(5), (May 1980):
    181-5. Cahners Publishing Co., 221 Columbus Avenue,
    Boston, MA 02116.
       Description of color graphics in computer use. In-
    cludes hardware requirements and needed managerial
    skills for rapid response to this emerging information
    technology.

Haggerty, Alfred G. "Pru Official Details Challenge of Full
    Financial Services," National Underwriter (Life/Health),
    8(47), (November 21, 1981):10-11. 420 East 4th Street,
    Cincinnati, OH 45202.
       Predictions suggest that by 1990 interactive television
    systems will be in one-third of all U. S. households.
    This trend will have implications for all sorts of finan-
    cial services, including insurance.

"Interactive Television: Telidon Brings the Wired City One
    Step Closer," Planned Innovation, 3(1), (January/Febru-
    ary 1980):5-6. NPN Infolink Limited, Management House,
    Parker T., London WC2B 5PU, England.
       A Canadian company, Telidon, has developed a new
    system to provide library facilities and graphics via
    television; the system is described.

Kleinberg, Ellen. "Interactive Disks: How Video Will
    Change the Sale," Industrial Marketing, 66(4), (April
    1981):46-49. Crain Communications Inc., 740 North
    Rush Street, Chicago, IL 60611.
       Description of possible uses for interactive video in
    the travel industry, at the Metropolitan Museum of Art,
    and other applications. Many possibilities for use can
    be explored.

Kraft, Robert N.; Dennis M. Buede; and John F. Patterson.
    The Design of a Multi-media Map-store/Surrogate Travel
    Information System, 1981. National Telecommunications
    Information Services, 608 13th Street, N.W., Washington,
    DC 20004.

Report of activities to design software and hardware
capabilities of a videodisc mapping and surrogate travel
system. Activities included selection of sites. Experts
from DARPA and ITC involved.

Louis, Arthur M. "Growth Gets Harder at Gannett, " Fortune,
103(8), (April 20, 1981):118+ . Time Inc. , Time and
Life Building, 1271 Avenue of the Americas, New York,
NY 10020.
Gannett Company, the nation's largest newspaper chain,
is considering many ways to utilize television transmission.

McCarthy, Patricia K. "Interactive Television: A Direct
Marketing Tool, " Direct Marketing, 42(10), (February
1980):30-48. Direct Marketing News Digest, 708 Silver
Springs Road, Rolling Hills Estates, CA 90274.
Description of three interactive systems that will be
utilized in interactive television. These systems arc
Prestel-Viewdata, QUBE, and The Source.

Mather, Boris. "The Electronic Working Environment, "
1980. (ERIC Document Reproduction Service Number
ED 192 393)
A review of the microelectronic equipment utilized in
the Canadian telephone industry indicating the impact
technology is having on the work force. Two-way interactive computer response systems are included in this
discussion.

Mazur, D. G. ; R. J. Mackey, Jr. ; S. G. Tanner; F. J.
Altman; and J. J. Nicholas, Jr. Forty and 80 GHz
Technology Assessment and Forecast Including Executive
Summary, 1976. National Telecommunications Information
Service, 608 13th Street, N. W. , Washington, DC 20004.
Results of a survey to determine current demands and
forecast growth in the demand for use of the 40 and 80
GHz bands during the period 1980-2000. Projected serv-
ices for interactive television are given with system re-
quirements and up and downlink calculations.

O'Brien, Terry, and Valerie Dugdale. "Questionnaire Administration by Computer, " Journal of the Marketing Research Society, 20(4), (October 1978):228-237. American Marketing Association, 250 South Wacker Drive,
Chicago, IL 60606.
Comparison of two methods of survey. In one method,
a computer-controlled television monitor and keyboard

were used to present questions and allow respondents to
record answers. In the second method, a conventional
field interviewer obtained information. The computer
response system had several advantages in cost efficiency
and "honesty" of answers.

O'Donnell, Thomas. "The Tube, The Ticker, and Jim
   Robinson," Forbes, 127(11), (May 25, 1981). 60 Fifth
   Avenue, New York, NY 10011.
       Report on the Columbus, Ohio QUBE two-way tele-
   vision system. Provides information on expanded services in banking and customer services available for
   purchasing at home through two-way television.

Pal, G. The Approaching Information Revolution and Its
   Possible Implications for Resource Sharing in Canada,
   Proceedings of the Seventh Annual Canadian Conference
   on Information Science, Banff, Alberta, Canada, Canadian Association on Information Science, May 12-15,
   1979.
       Discussion of the national telecommunications system
   and the potential growth of information sharing. Interactive television could be utilized by libraries to provide resource sharing programs.

A Planning Study to Develop a Demonstration for the Use of
   Telecommunications in Public Service Delivery. Final
   Report, Department of Health, Education, and Welfare,
   Washington, D. C. , June 30, 1974. (ERIC Document
   Reproduction Service Number ED 100 317)
       Telecommunications equipment of the mid-1970's was
   surveyed to ascertain delivery of public services via tele-
   communications. Various applications were proposed.
   An appendix analyzes the cost and hardware requirements.

Reiss, Craig. "The Dark Side of Cable," Marketing and
   Media Decisions, 16(8), (August 1981):62-63, 118, 120.
   342 Madison Avenue, New York, NY 10017.
       The possible misuse of data collected through inter-
   active television systems is discussed.

Rush, Ramona R. "The Teletexts, Videotexts, Ceefaxs Are
   Coming," Journal of Organizational Communication, 9(30)
   (Second Quarter, 1980):3-5. International Association of
   Business Communicators, 870 Market Street, Suite 940,
   San Francisco, CA 94102.
       A description of various formats and uses for information

dissemination into homes, as the interactive television capability becomes more widespread.

Spak, G. T. How to Operate an Energy Advisory Service, Volume I, Report and Recommendations, March 1978. New York Institute of Technology, Center for Energy and Policy and Research, Northern Boulevard, Old Westbury, NY 11568.
Various channels of communication were used to disseminate energy conservation information. Professionals concerned with energy conservation were supplied information via innovative interactive television seminars. Follow-up survey indicated a positive response to the dissemination of energy information, with positive results.

Stetten, Kenneth J. Interactive Television Software for Cable Television Application, June 1971. Mitre Corporation, McLean, VA (ERIC Document Reproduction Service Number ED 056 522)
The premise is considered that if the content of cable television services were broadened to provide interactive communication, people might adapt more easily to large urban communities, using cable to provide new communication links. TICCIT (Time-Shared, Interactive, Computer-Controlled Information Television) is described in the appendix.

Survey of Two-Way Cable Television Testbeds, May 1974. Cable Television Information Center, Washington, DC. (ERIC Document Reproduction Service Number ED 095 894)
Surveys of 10 two-way interactive cable experiments prior to 1974. Parent companies, locations of test beds, and descriptions of each project are included.

Tague, J., and F. Dolan. Teletext: Its Development and Social Implication, Proceedings of the Seventh Annual Canadian Conference on Information Science, Banff, Canada, Canadian Association of Information Science, May 12-15, 1979.
Discussion of the development of teletext systems in several countries, with emphasis on the systems used in Canada. Benefits, problems, and utilizations of teletext systems with interactive television are indicated.

Teletext and Public Broadcasting: A Report Prepared for the Corporation for Public Broadcasting. April 1980. National

Telecommunications and Information Administration,
Alternate Media Center, School of the Arts, New York
University, New York, NY 10003.
    Report of a planning study that reviewed potential
applications of teletext within public broadcasting. Includes the steps needed to implement teletext services.

Test and Evaluation of Public Service Uses of Cable Tele-
    vision: Reading, Pennsylvania, 1976. National Science
Foundation, Washington, DC, RANN Program. (ERIC
Document Reproduction Service Number ED 125 548)
    The New York University-Reading Consortium pro-
posed to evaluate the benefits of delivering public services to a segment of elderly population through a system
of interconnected neighborhood community centers. Costs,
social benefits, and impact of the interconnected system
were to be examined.

Thomas, Willard. "Interactive Video," Instructional Innovator,
    26(2), (February 1981). Assocation for Educational Com-
munication and Technology, 1126 Sixteenth St., N.W.,
Washington, DC 20036.
    Examination of the instructional procedures needed in
producing interactive video programs. Traditional instructional materials, which do not allow for interruption
and replay, are counterproductive in the use of the interactive, branching capability available with interactive television. The full value of interactive hardware is lessened
if traditional software is utilized.

Veith, Richard H. A Survey of Two-way Cable Television
    Systems, Including Field Experiments and Pilot Programs,
January 1973. Master of Arts Thesis submitted to
California State University, San Francisco, CA.
    Several experimental television systems, located
primarily in California, were described and analyzed in
the development of this graduate study.

Volk, John L. "Interactive Television Experiment in Reston,
    Virginia," Journal of Educational Technology Systems,
3(1), (Summer 1974):73-84. (Formerly entitled Journal
of Educational Instrumentation.) Baywood Publishing
Company, Inc., 120 Marine Street, Box D, Farmingdale,
NY 11735.
    Describes the early 1974 Reston, Virginia, interactive
television system and discusses future plans.

Volk, John. The Reston, Virginia Test of the Mitre Corpo-

ration's Interactive Television System, May 1971. Mitre
Corporation, 1820 Dolly Madison Blvd., McLean, VA
22102. (ERIC Document Reproduction Service Number
ED 056 523)
    Description of TICCIT (Time-Shared Interactive Com-
puter-Controlled Information Television) as it would be
used with the home terminal. A demonstration software
package for education and community service is included.

Wollman, Jane. "Video As a Sales Tool," Output, 2(2),
(April 1981):28-32. Output Magazine, Technical Publishing, 875 Third Avenue, New York, NY 10022.
    Description of the uses of video as a sales and adver-
tising tool. Includes mention of interactive video, which
involves consumers in entering and extracting data. Two-
way television for shopping, teletext, silent commercials,
video disc, and special interest broadcasting is discussed.

Wood, Fred B., and others. "Videoconferencing Via Satellite: Opening Government to the People," The Futurist:
A Journal of Forecasts, Trends, and Ideas About the
Future, 12(5), (October 1978):321, 323-326. World
Future Society, 4916 St. Elmo Avenue, Bethesda, MD
20014.
    Satellite videoconferencing is technically feasible. The
U. S. Congress could begin to utilize technology to
strengthen communication and understanding between citizens and congressmen more efficiently.

Woolf, Thomas M. "Satellite Communication for Communication Law," Educational and Industrial Television, April
1980. C. S. Tepfer Publishing Company Inc., 51 Sugar
Hollow Road, Danbury, CT 06810.
    Law students in California studied communications law
under a professor of law at the School of Law of New
York University. Interactive communication occurred
through PSSC, Westor I, and AT&T dedicated telephone
lines.

Yamaguchi, K., and others. "Coaxial Cable Information
System with Interactive Television Services," Two-way
Cable Television, 11(292), (April 1977):121-31. Springer
Verlag Publishing Company, 175 Fifth Avenue, New York,
NY 10017.
    Description of an interactive cable television system
developed in the suburbs of Tokyo. This experimental
program became operational in 1976.

EDUCATION

Abt, Clark C. What The Future Holds for Children in the
     TV Computer Age: Unprecedented Promises and Intoler-
     able Threats to Child Development. Keynote address to
     the National Council for Children and Television Sympos-
     ium on Children, Families and New Video/Computer
     Technologies. Princeton, New Jersey, March 10, 1980.
     (ERIC Document Reproduction Service Number ED 194
     215)
          This address describes video communication technology
     and computer technology combined in a new form of inter-
     active television. Simultaneous program capacity and
     manipulation of program materials is possible, at costs
     and prices accessible to 80 percent of the American and
     European households. Positive and negative results of
     this technological advancement for children are listed.
     The author takes the position that government and industry
     each have an obligation to inform consumer parents and
     children of the impact of this technology. Methods of
     use, quantity of use, age of participants, and human cap-
     abilities are considered.

Allen, Brockenbrough S. "The Video-computer Nexus:
     Towards an Agenda for Instructional Development,"
     Journal of Educational Technology Systems, 10(2), (1981-
     82):81-99. (Formerly entitled Journal of Educational
     Instrumentation.) Baywood Publishing Company, Inc.,
     120 Marine Street, Box D, Farmingdale, NY 11735.
          Integration of video and computer technologies will
     have a powerful influence on products in the education
     and entertainment fields. Factors that will influence
     social and economic values in this combined technology
     are outlined. The value of instructional theory for prop-
     er application is noted.

American Indian Telecommunications Satellite Demonstration
     Project Summary Report, May 1979. All Indian Pueblo
     Council, Albuquerque, New Mexico. Bureau of Indian
     Affairs (Department of Interior), National Aeronautics
     and Space Administration, Washington, DC. (ERIC
     Document Reproduction Service Number ED 173 012)
          Report of a two-way interactive television conference
     in April 1978 between American Indian tribes and gov-
     ernment agents. Locations in Montana, New Mexico,
     California, and Washington, D. C., allowed discussion

and question-and-answer sessions. The feasibility of
satellite communication to improve the Indian Information
Network was part of the content and discussion.

Anandam, Kamala. Annual Report for Open College, 1976-77,
1977. Miami-Dade Community College, FL 33156.
(ERIC Document Reproduction Service Number ED 156
272)
     Report on a computer-based feedback system for
individualized instruction, Response System with Variable
Response (RSVR). The program is used with adult students who have difficulty attending classes on campus,
but feel a need for classroom sessions.

Andrissen, J. J., and D. J. Kroon. "Individualized Learning by Videodisc, " Educational Technology, March 1980.
Educational Technology Publications, 140 Sylvan Avenue,
Englewood Cliffs, NJ 07632.
     Report of a pilot project using a specially construct-
ed VLP (Video Long Play) videodisc operating under the
control of a minicomputer. System description, observations, and recommendations for future computer interactive VLP courses are presented.

The Appalachian Education Satellite Project Executive Report,
     1976. Appalachian Education Satellite Project, Lexington,
Kentucky, Department of Health, Education, and Welfare,
Washington, DC. (ERIC Document Reproduction Service
Number ED 125 594).
     Report of the ATS-6 Satellite communication project
of the early 1970's transmitting into the Appalachian area
of the United States.

Blackburn, Gary M. Colloquium Series on Career Education
for Handicapped Adolescents 1976, 1976. Purdue University, Special Education Section from the Department
of Education, West Lafayette, IN 47907. (ERIC Document Reproduction Service Number ED 143 791)
     Texts from videotaped presentations that were part of
the project "Interactive Television Colloquium Series on
Career Education for Handicapped Adolescents" are
presented in this document.

Bloch, David C. "How to Plug into Your Umbilical Cord, "
Instructional Innovator, 26(2), (February 1981). Association for Educational Communication and Technology,
1126 Sixteenth Street, N.W., Washington, DC 20036.

Review of cable television systems. Brief descriptions indicate methods to utilize closed-circuit television, including two-way interaction within educational institutions.

Bork, Alfred. "Educational Technology and the Future," Journal of Educational Technology Systems, 10(1), (1981-82):3-20. Baywood Publishing Company, Inc., 120 Marine Street, Box D, Farmingdale, NY 11735.
    An outline of future uses of computers and optical videodiscs in education. Implications of this merger of technology are discussed.

Bramble, William J., and Claudine Ausness. An Experiment in Educational Technology: An Overview of the Appalachian Education Satellite Program, August 1974. Department of Health, Education, and Welfare, Washington, DC. (ERIC Document Reproduction Service Number ED 103 007)
    Examination of the ATS-6 Satellite program conducted in the Appalachian area of the United States. The efficiency of several methods of return communication is discussed.

Carey, John. "A Primer of Interactive Television," Journal of the University Film Association, 30(2), (Spring 1978). University Film Association, Department of Cinema and Photography, Southern Illinois University, Carbondale, IL 62901.
    A description of four major components of television programming: time, space, visual conventions, and participants. These components are examined, contrasting the characteristics of commercial (often pre-taped and edited) programs and interactive, live television. The two formats of television do not have identical components; indeed, the differences in production and taping create very different content and programming.

Carpenter, Robert L. "Closed Circuit Interactive Television and Inservice Training," Exceptional Children, 45(4), (January 1979). The Council for Exceptional Children, 1920 Association Drive, Reston, VA 22091.
    The use of closed-circuit interactive television in teacher in-service training, especially applicable for teaching handicapped children, is described.

Carpenter, Robert L. "Multipoint Closed Circuit Interactive Television As a Response Mode to the Need for In-service

Training in Special Education, " Educational Technology, 18(11), (November 1978):16-19. Educational Technology Publications, 140 Sylvan Avenue, Englewood Cliffs, NJ 07632.
   A description of an interactive television system developed for in-service with handicapped professionals in the Indiana Higher Education Telecommunications System. A teleresponse system to remote studios was used. A model of the network is included.

Cerf, V. G. , and others. Creation of a Testbed for Innovative Educational TV Experiments, June, 1975. Final Report, National Science Foundation, Washington, DC and Stanford University, California, Center for Interdisciplinary Research. (ERIC Document Reproduction Service Number ED 176 999)
   Report on two interactive television systems built at Stanford University. Digital systems are employed in both systems, which are intended to serve as "testbeds" to develop new instructional technologies.

Charles, Joel. "Interactive Television: Use Your Imagination and Help Students Learn, " Audiovisual Instruction, 21(5), (May 1976):23-25. (Currently entitled Instructional Innovator.) Association for Educational Communication and Technology, 1126 Sixteenth Street, N.W. , Washington, DC 20036.
   There are indications that interactive television instruction elicits psychomotor responses by students during the teaching process.

Curtis, John A. , and Joseph M. Biedenbach. Educational Telecommunications Delivery Systems, 1976. The American Society for Engineering Education, One Dupont Circle, Suite 400, Washington, DC 20036.
   Cable television, radio, and ITFS are described through applications relating to educational uses. Adult education applications, for credit and noncredit courses, as well as services to the handicapped and high school applications are examined.

Curtis, John A. , and Alan H. Blatecky. "The Economics of ITFS Use: An In-Depth Study of Selected Systems by Project TIMES, " Educational and Industrial Television (E&ITV), 10(4), (April 1978):47-58. C. S. Tepfer Publishing Company, Inc. , 51 Sugar Hollow Road, Danbury, CT 06810.

A review of the value of ITFS as an educational tool.
Includes a report of the re-study of 19 schools (11 public, 4 private, and 4 higher education) examined in the
Project TIMES study conducted by the Center for Excellence, Inc. during 1975-76.

"The Cutting Edge: Toronto Plugs into Huge Videotext Utility," American Libraries, 13(4), (April 1982):266.
American Library Association, 50 East Huron Street,
Chicago, IL 60611.
Fifteen hundred computer terminals will be available
for citizens in Toronto; located in libraries, shopping
centers, and public institutions, citizens may access
several levels of information.   "Teleguide to Ontario"
is completely interactive; computer terminology is easily understood by Canadian users.

Dechenne, James A. "Talkback Television: A Partial Solution to Declining Enrollment," Educational Technology,
20(10), (October 1980). Educational Technology Publications, 140 Sylvan Avenue, Englewood Cliffs, NJ 07632.
South Oklahoma City Junior College (SOCJC), as a
member of the Oklahoma Higher Education Television
System, participates in a Talkback Television program.
TBTV provides two-way audio communication between
students and teachers at the college level.

Depover, C. "An Interactive Student Interrogation System
Using a Microcomputer," Educational Technology, 22(1),
(February 1982):25-26. Educational Technology Publi-
cations, 140 Sylvan Avenue, Englewood Cliffs, NJ 07632.
A description of a microcomputer program in use at
the University of Mons, Belgium.

Doyle, Brian. "Interactive Video: The Osseo Project,"
Minnesota Media, June 1983, pp. 10-11. Ray Birr,
President Executive Committee, MEMO (Minnesota
Educational Media Organization), 17870 Italy Path, Lakeville, MN 55044.
Osseo (Minnesota) Area Schools developed teaching
units using videodisc and microcomputers to create interactive video.   The project was funded through the Minnesota legislature; pilot programs employed laser videodisc and videotape with microcomputers.   This article
provides details of development, subject areas, cost,
evaluation of products, and further interactive projects.

Fruchter, Dorothy A., and George M. Higginson. An Evaluation Report on Project Interact: A Teacher In-service
Training Course on Career Education Using Two-way TV
in Texas to Several Groups Simultaneously, December
1975. (ERIC Document Reproduction Service Number ED
110 076)
    A third-party evaluation was conducted of an experi-
ment using two-way television to reach several groups
simultaneously. Audiences were teachers involved with
an in-service training program in career education. A
Texas Career Education Grid was employed. The results indicated that staff dedication was high, but mechanical failure of the Grid impaired the program.

Gayeski, Diana M., and David V. Williams. "Program
Design for Interactive Video," Educational and Industrial
Television (E&ITV), 12(12), (December 1980):31-34. C.
S. Tepfer Publishing Company, Inc., 51 Sugar Hollow
Road, Danbury, CT 06810.
    An examination of the design considerations needed to
increase instructional value through the interactive video
format. Commentary on hardware components and the
need for high quality multiple choice questions is included.

Glenn, Allen D., and Kent T. Kehrberg. "The Intelligent
Videodisc: An Instructional Tool for the Classroom,"
Educational Technology, 21(10), (October 1981). C. S.
Tepfer Publishing Company, Inc., 140 Sylvan Avenue,
Englewood Cliffs, NJ 06810.
    The MECC-Rockefeller Family Fund Project has
linked the Pioneer Model VP-1000 optical videodisc with
the Apple II computer. Thirteen learning sessions in
economics are undergoing pilot tests in selected schools.
The potential advantages and key issues in economic hard-
ware selection are discussed.

Goldberg, Leonard M., and Frank S. Rubin. "Interactive
Computer-controlled TV for the Deaf," Audiovisual Instruction, January 1978. (Now entitled Instructional
Innovator.) Association for Educational Communication
and Technology, 1126 Sixteenth Street, N.W., Washington, DC 20036.
    A description of the interactive computer delivery sys-
tem developed for educational experimentation at the
Model Secondary School for the Deaf in Washington, D.C.
The Mitre Corporation combined standard TICCIT (Time-
shared, Computer-Controlled Information Television)

features with specially designed components using a
larger computer, creating the TICCIT + 10 system.

Goldman, Ronald J. Demand for Telecommunications Services in the Home, May 1979. (ERIC Document Reproduction Service Number ED 177 616)
    A paper presented at the Annual Meeting of the Inter-
national Communication Association in Philadelphia, Pennsylvania, May 1-5, 1979. The paper includes a list of
117 services potentially available via telecommunications.

Green, David, and Bill Lazarus. "A User-Controlled Teleconference Studio," Educational and Industrial Television
(E&ITV), April 1980. C. S. Tepfer Publishing Company,
Inc., 51 Sugar Hollow Road, Danbury, CT 06810.
    A description of the four-channel ITFS system operated
by the Educational Television Center of the Archdiocese
of San Francisco. Teleconferencing has been used in a
number of ways, including live interactive dance perform-
ances. The Educational Television Center transmission
signal covers a large part of northern California.

Greene, Alexis. "Poor Ratings for Two-way Television,"
Change: The Magazine of Higher Learning, 7(4), (May-
June 1979):56-57. Heldref Publications, 4000 Albermarle
Street, N. W., Washington, DC 20016.
    Information concerned with college courses offered by
the Higher Education Cable Council (HECC) and Warner
Cable Corporation through the QUBE interactive tele-
vision system in Ohio.

Harris, Marilyn, and Melodee Williams. "Television Comes
to the Classroom in Irvine--Anyone Can Become a Star,"
Thrust for Educational Leadership, 8(1), (October 1978).
Association for California School Administrators, 1575
Old Bayshore Highway, Burlingame, CA 94010.
    An interactive cable television network in the Irvine,
California school district permits communication between
students and teachers in different schools. The community cable television provides viewing of school programs
for cable subscribers.

A High Power Primer on Low Power Television, 1981.
EMCEE Broadcast Products, Inc., Post Office Box 68,
White Haven, PA 18661.
    This booklet presents a review of the Federal Commun-
ication Commission Notice of Proposed Rulemaking regarding low-power television.

How Now Brown Cow? The Texas Educational Telecommunications Study, 1975. Educational Development Corporation,
2813 Rio Grande, Austin, TX 78705. (ERIC Document Reproduction Service Number ED 110 052)
A review of the media to be used across Texas for educational instructional services, particularly within the public schools.

Hoye, Robert E. Methodology and Models of Learning, November 1976. (ERIC Document Reproduction Service Number
ED 125 576)
This publication indicates that the entire educational
system needs to be considered in order to make the best
use of modern technology to improve learning.

Hubin, Allen J. Video Network Instruction: A Critical View,
November 1978. (ERIC Document Reproduction Service
Number ED 163 547)
A paper presented at the Annual Meeting of the Speech
Communication Association in Minneapolis, Minnesota,
November 2-5, 1978. Presents a review of UNITE (Univer-
sity-Industry Television for Education), the system oper-
ated by the University of Minnesota. The Talk-back Tele-
vision Classroom feature of the system is one component
discussed.

ITFS in America Today: The EMCEE Roundtable, 1(1), (Feb-
ruary 1982). Published by EMCEE Broadcast Products,
Inc., Post Office Box 68, White Haven, PA 18661.
A description of four ITFS systems operating in the
United States within educational and medical institutions.

Institute for Research and Development in Occupational Education
July 1, 1977 - June 30, 1978. Annual Report Number 7,
September 1978. City University of New York, New York
Institute for Research and Development in Occupational
Education, New York State Education Department, Albany.
Division of Occupational Education Supervision. (ERIC
Document Reproduction Service Number ED 171 888)
Presents an overview of the major projects completed
in occupational education, including a project incorporating
two-way television.

Katz, Lawrence S. "Potentials of Interactive Cable Television, "
in Cora E. Thomassen, ed. , CATV and Its Implications
for Libraries (University of Illinois, Graduate School of
Library Science, Urbana-Champaign, Illinois, 1974).

An overview of the emerging technologies developed by
the Mitre Corporation and the potential applications in providing information services via television. Description of
the TICCIT (Time-shared, Interactive, Computer-Controlled Information Television) system, which can provide
interactive communication for education, community, and
health services.

Kenney, Brigitte L., and Frank W. Norwood. "CATV: Visual Library Service," American Libraries, 2(6), (July/
August 1971):723-26. American Library Association,
50 East Huron Street, Chicago, IL 60611.
A Visual Reference Service provided by the Natrona
County, Wyoming, Public Library is described. This
pilot program indicates some of the potential applications
for interactive television services for the community.

Kirman, Joseph M., and Jack Golberg. "Distance Education:
Teacher Training via Live and Concurrent Group Telephone Conferencing," Educational Technology, 21(4),
(April 1981):41-42. Educational Technology Publications,
140 Sylvan Avenue, Englewood Cliffs, NJ 07632.
One-way television combined with group telephone conferencing appears to be as effective as face-to-face instruction. The variety of telecommunication formats and
the required planning in each installation are discussed.

Kitchen, Will. "Microwave Television and Interactive Cable,"
Minnesota Media, June 1983, pp. 13-15. Ray Birr,
President Executive Committee MEMO (Minnesota Edu-
cational Media Organization), 17870 Italy Path, Lakeville,
MN 55044.
A review of the interactive television network under
development among ten school districts in east central
Minnesota. Cooperation and support of the Minnesota
Department of Education plus anticipated financial sup-
port from the Bush Foundation and the U. S. Department
of Commerce National Telecommunications and Information
Agency will assist in the development of this program.

Larimer, George S., and W. Ward Sinclair. "Some Effects
of Two-way Television on Social Interaction," AV Communication Review, 17(1), (Spring 1969). (Currently entitled ECTJ: Educational Communication and Technology)
Association for Educational Communication and Technology,
1126 Sixteenth Street, N. W., Washington, DC 20036.
A report of research in two-way television, supported

by the Central Fund for Research, Pennsylvania State
University, Grant Number 151, State College, PA 16801.

Nocerino, Joseph T., and others. Interactive Television:
A Pathway to Futures in Special Education Using Telecommunications, June 1978. Paper presented at the
First World Congress on Future Special Education,
Stirling, Scotland, June 25-July 1, 1978. (ERIC Document Reproduction Service Number ED 157 334)
This paper describes the combination of a computer-
assisted instructional system, a cable television system,
and multimedia equipment to provide instructional sessions. The primary audience is the handicapped child,
in home or school meetings.

Norwood, Frank W. "Education by Satellite in the USA, "
Educational Media International, 2, (1981):19-22. Edu-
cational Media Yearbook, Librarioo Unlimited Inc.,
Post Office Box 263, Littleton, CO 80160.
A review of experimental satellites employed during
the 1970's over the North American continent.

Orlich, Donald C., and Charles E. Hllen. Telecommunications for Learning, 1969. (ERIC Document Reproduction
Service Number ED 031 963)
A report of a seminar on applications of telecommunications technology to educational theory; the report contains eleven speeches. Speech number ten is concerned
with the uses of two-way television in the future of
education.

Park, Ben Kimball. The Influence of the Structural Characteristics of the Television Image on Human Communication in Interactive Television, 1978. Unpublished doctoral dissertation, New York University, 905 Tisch Hall,
New York, NY 10003.
Report of a study conducted at Reading, Pennsylvania.
The study identified factors that created difficulty in
communication between participants at locations involved
in interactive television transactions.

Parker, Lorne A., and Marcia A. Baird. "Humanizing
Telephone-Based Instructional Programs, " University of
Wisconsin-Madison, University-Extension, November
1975. (ERIC Document Reproduction Service Number
ED 125 592)
Text presents specific reports on many teleconferencing

projects in the fields of education and medicine. Applications utilizing audio communication, and audio and video, one- and two-channel return systems are included.

Pittman, Theda Sue, and James Orvik. ATS-6 and State Telecommunications Policy for Rural Alaska: An Analysis of Recommendations, December 1976. Alaska University, Center for Northern Educational Research, Fairbanks, AK 99701. (ERIC Document Reproduction Service Number ED 142 170)

This paper analyzes thirteen recommendations for media policy-making in rural Alaska which were formulated as a result of the Appalachians Technology Satellite project in 1974-75. A brief description of the Educational Satellite Communications Demonstration/Alaska project is provided, plus a map of receiving sites. Information regarding transmission capabilities and a bibliography of additional sources of information for related ATS-6 projects is also included.

Richardson, Mary Ellen. Parents Role in Using Interactive Television with Handicapped Children, 1981. Unpublished doctoral dissertation, State University of New York College at Buffalo, Kapen Hall, Buffalo, NY 14260.

Behaviors of parents with handicapped children were observed in this study. Students were provided learning activities via interactive television. A descriptive investigation/interview with parents indicated a positive response to the use of interactive television in the home.

Rimmer, Tony. Viewdata--Interactive Television, with Particular Emphasis on the British Post Office's Prestel, 1979. (ERIC Document Reproduction Service Number ED 188 214)

An overview of Viewdata is presented, including technical aspects and problems of introducing the system into Great Britain. The relevance of this information to the adoption of a viewtext system in the United States suggests that delay would be advantageous.

Showalter, Robert G. Purdue Interactive Television Colloquium Series: Continuing Career Education via IHETS Television Network Final Report, 1975. Bureau of Education for the Handicapped (DHEW/OE), Washington, DC (ERIC Document Reproduction Service Number ED 144 335)

A report of a two-year prototype project which used multipoint closed-circuit interactive television to provide career education to specialists working with language-disordered children.

Stetten, Kenneth J. TICCIT: A Delivery System Designed for Mass Utilization, 1977. (ERIC Document Reproduction Service Number ED 056 525)
Description of TICCIT (Time-shared, Interactive, Computer-Controlled, Information Television), a system which will allow low-cost delivery of computer services to homes and schools through interactive television. Minicomputer facilities, hardware problems, on-line terminals, TICCIT software, and efficient instruction-writing languages are discussed.

A Study of a Proposed Multi-purpose Communications System, 1971. (ERIC Document Reproduction Service Number ED 055 453)
A faculty committee composed of various members from Washington State University reported on the feasibility of establishing a two-way television network in the southeastern area of the State of Washington. This network would eventually cover Washington, and would include voice communication, teletype, and facsimile reproduction.

von Feldt, James R. Description of a Computer-based, Interactive, Instructional Television System Prototype, 1977. (ERIC Document Reproduction Service Number ED 160 086)
The potential value of computerized, interactive television systems is discussed. Emphasis is upon the instruction-demanding motion, with concern for the most effective and efficient production technique to encourage learning.

von Feldt, James R. A Description of the DAVID Interactive Instructional Television System and Its Application to Post High School Education of Deaf, December 1978. National Institute for the Deaf, Rochester Institute of Technology, Rochester, NY 14623.
This paper describes the interactive, instructional television system DAVID (Digital And Video Interactive Device) developed at the National Institute for the Deaf by Dr. von Feldt. The information included delineates the student audience, completed prototype equipment, and the results of preliminary testing of the system with a specific audience.

Wall, Shavaun M., and Nancy E. Taylor. "Using Interactive
    Computer Programs in Teaching Higher Conceptual Skills:
    An Approach to Instruction in Writing," Educational Tech-
    nology, 22(2), (February 1982):13-17. Educational Tech-
    nology Publications, 140 Sylvan Avenue, Englewood Cliffs,
    NJ 07632.
        Presentation of an approach to the use of computers
    for instruction in descriptive writing through an inter-
    active framework.

Wellens, A. R. "A Device That Provides an Eye-to-eye
    Video Perspective for Interactive Television," Behavior
    Research Methods and Instrumentation, 10(1), (February
    1978):25-6. Psychonomic Society, Inc., 2904 Guadalupe
    Street, Austin, TX 78705.
        Description of a device that allows individuals to en-
    gage in eye contact while interacting over two-way tele-
    vision. This equipment would be useful for observing
    the non-verbal communication patterns of individuals en-
    gaged in social interaction.

Wellens, A. Rodney. "An Interactive Television Laboratory
    for the Study of Social Interaction," Journal of Non-ver-
    bal Behavior, 4(2), (Winter 1979):119-122. Human
    Sciences Press (subsidiary to Behavioral Publications),
    72 Fifth Avenue, New York, NY 10011.
        Description of a device that allows individuals to make
    eye contact while interacting through two-way television.

West, Peter C. A Survey and Report of Interest in and Avail-
    ability of Systems for the Delivery of Instruction by Re-
    mote Methods, 1980. (ERIC Document Reproduction
    Service Number ED 192 695)
        An examination of the feasibility of providing upper-
    division and graduate-level instruction in the Rockford,
    Illinois, area by "remote delivery."

Weston, J. R. Teleconferencing and Social Negentropy,
    1973. (ERIC Document Reproduction Service Number
    ED 078 480)
        Presents models of closed-circuit television systems
    that allow self and interpersonal interactions and role
    perceptions to develop. The article indicates that elec-
    tronic transmission communication can be a positive
    force in establishing interpersonal communication.

Wood, Fred B., and others. "Videoconferencing via Satellite:

Opening Congress to the People, Summary Report," 1978. National Aeronautics and Space Administration, Lewis Research Center, Cleveland, OH. (ERIC Document Reproduction Service Number ED 162 647)
A description of personal evaluations by United States congressmen of satellite videoconferencing with two-way interactive television between congressmen and constituents. Seven questions were used to elicit information about this proposed communication procedure.

Zenaty, Jane, and others. "A Minicomputer Software System for Administering Interactive Instructional Programs via Two-way Cable Television," Journal of Computer-based Instruction, 5(4), (May 1979). Association for Development of Computer Based Instructional Systems, Western Washington University, Bellingham, WA 98225.
The article describes a minicomputer software system combining interactive cable television and computer-aided instruction. Initial application was for the administration of instructional programs for potentially large populations.

Zenaty, J. W., and others. "A Minicomputer Software System for Two-way Cable Television. Computer Assisted Instruction on Channel 4, Supermarket Shopping on Channel 7, Mail Delivery on Channel 10," Proceedings of Spring Computer Conference, 1978, San Francisco, California. In IEEE (Institute of Electrical and Electronic Engineers), XVII, (February 28/March 3, 1978). Contact Institute of Electrical and Electronic Engineers, Inc., 345 East 47th Street, New York, NY 10017.
Description of a two-way cable television demonstration experiment that provided computer-aided instruction. The data indicated that large numbers of students could be served at a relatively low cost.

Zraket, Charles A. "Some Technical, Economic, and Application Considerations of Interactive Television," January 1974. National Science Foundation, 1800 G Street, N.W., Washington, DC 20006 (ERIC Document Reproduction Service Number ED 081 220)
A National Science Foundation Report about the changed image required for CATV (Cable Access Television) in order for CATV to provide new and expanded communication services. These services could be provided through interactive television technology.

MEDICINE

Andrus, W. S.; J. R. Dreyfus; F. Jaffer; and K. T. Bird.
"Interpretation of Roentgenograms via Interactive Television, " Radiology, 116(1), (July 1975):25-31.   Radiological Society of North America, Inc. , Oak Brook Regency
Towers, 1415 West 22nd Street, Suite 1150, Oak Brook,
IL 60521.
    Report on the examination of roentgenograms (photos
of x-rays) by interactive television and by direct viewing.
The results indicated that interpretations through television were of an acceptable accuracy.

ATS-6 Health Experiment: Indian Health Service/Alaska
    WAMI Experiment in Regionalized Medical Education/
Seattle, Washington, Phase II: Operations, 1975.   National Library of Medicine, 8600 Rockville Pike, Bethesda,
MD 20209.
    A review of the operational phase of the ATS-6 Health
Education Telecommunications Experiments.   Interactive
video, audio, and data communications linked faculty at
the University of Washington with faculty at the University of Alaska and the Community Clinical Unit at Omak,
Washington.

Bashshur, Rashid; Patricia A. Armstrong; and Zakhour I.
    Youseff. Telemedicine: Explorations in the Use of
Telecommunications in Health Care, 1975.   Conference
Workshop, Ann Arbor, Michigan, October 29-30, 1973.
Published by Charles C. Thomas, 2600 South First,
Springfield, IL 62704.
    Report of a project which investigated the efficiency,
impact, and acceptance of telemedicine for the delivery
of health services in a rural setting.

Baxter, W. Eugene. Substitute System for Rural Health
    Care, 1976.   National Technical Information Service,
Report Number HRP-0015608/3SL, Springfield, VA 22161.
    A report on the use of two-way television systems to
provide backup information for physician extenders.
Photographs of equipment, details of staffing arrangements, and copies of protocols for some systems are
included.   The recommended design features direct communication between the patient and the physician through
interactive television.

Bergen, Bernard J. , and others. Psychiatric Consultation

via Television, 1972. Final Report, National Institute of
Mental Health, Grant Number MH 15007. National Insti-
tute of Mental Health, Alcohol, Drug Abuse, and Mental
Health Administration, 5600 Fishers Lane, Rockville,
MD 20857.
   Report of a two-year study that provided psychiatric
consultation through closed circuit two-way television.
The intended audience was the non-psychiatric physician.
Continuing education was also provided to this select
audience.

Carpenter, Robert L. "Closed Circuit Interactive Television
and Inservice Training," Exceptional Children, 45(4),
(January 1979):289-90. The Council for Exceptional
Children, 1920 Association Drive, Reston, VA 22091.
   A discussion of closed-circuit interactive television in
teacher in-service training. A project involving handicapped students and career education is described.

Dwyer, Thomas F. "Interactive Television in Psychiatric
Interviewing and Treatment," Medical Tribune, 11(55),
(1970):8. Medical Tribune Inc., 641 Lexington Avenue,
New York, NY 10022.
   A description of a teleconsultation program connecting
Massachusetts General Hospital and the Bedford, Massachusetts Veterans Administration Hospital. The program
used the teleconsultation system for any activity that
could be communicated better by television than by telephone, mail, or physical presence.

Dwyer, Thomas F. "Telepsychiatry: Psychiatric Consultation
by Interactive Television," American Journal of Psychiatry, 130(8), (1973):865-869. American Psychiatric
Association, 1700 18th Street, N. W., Washington, DC
20009.
   A discussion of telepsychiatry (psychiatric consultation
by interactive television). The system developed between
Massachusetts General Hospital and Boston Medical Station
is described. Implications of the future use of telepsychiatry are included in the discussion.

Felton, Barbara J.; Mitchell Moss; and Rafael J. Sepulveda.
"Two-way Television: An Experiment in Interactive Pro-
gramming for the Elderly," Experimental Aging Research,
6(1), (1980):29-44. Beech Hill Publishing Company, P.O.
Box 29, Mount Desert, ME 04660.
   The report of a study conducted to assess the effectiveness

of a two-way cable television system in reaching older
people in Reading, Pennsylvania. The viewing frequency
of the intended audience was influenced by programming
local events with the focus upon senior citizens. The
cable system successfully reached the target audience
within two years.

Fry, Carlton F., and others. "Interactive Television in
    Nursing Continuing Education," Journal of Continuing
    Education in Nursing, 7(3), (May/June 1973). Charles
    B. Slack, Inc., 6900 Grove Road, Thorofare, NJ 08086.
        Describes the live color microwave television trans-
    mission system used to teach a course in critical care
    nursing; the system has two-way audio and video com-
    munication. This course is sent from an urban univer-
    sity medical center to staff members in rural southeast-
    ern Ohio hospitals.

Goldberg, Myron L., and A. Rodney Wellens. "A Compar-
    ison of Nonverbal Compensatory Behaviors Within Direct
    Face-to-face and Television-Mediated Interviews," Journal
    of Applied Social Psychology, 9(3), (1979):250-60. V. H.
    Winston and Sons, 7961 Eastern Avenue, Silver Springs,
    MD 20910.
        Direct face-to-face and two-way television mediated
    interviews were conducted. Looking, talking, smiling,
    and listening activities were timed and scored from video-
    tapes of the interviews. More looking than listening
    occurred in the two-way television interviews. No other
    differences were noted between the two types of inter-
    views. Results of this study as applicable to future
    social interactions via two-way television are discussed.

Hageboeck, Mary K., and others. Interactive Television: A
    Study of Its Effectiveness as a Medical Education Resource
    in the Rural Northwest, 1973. National Technical Infor-
    mation Service, Report Number PB-225 172/6, Springfield,
    VA 22161.
        A review of a model interactive television system serv-
    ing rural New England.

Hageboeck, Molly, and Leon J. Rosenberg. Practical Con-
    cepts for Using Interactive Television: Applications in
    the Medical Community, 1975. National Technical Infor-
    mation Service, Report Number PB-241 319/3SL, Spring-
    field, VA 22161.
        Practical Concepts Incorporated (PCI) was retained to

provide market analysis and management assistance in
support of an interactive television network serving the
medical community in rural New England. The CPI report indicated that the self-sufficiency of the system
must be considered from the onset of planning or the
system may not be considered successful.

Hoehn, Robert E., and John Givens. "Astroteaching," Audio-
    visual Instruction, 22(3), (March 1977):52-53. (Retitled
    Instructional Innovator.) Association for Educational Com-
    munications and Technology, 1126 Sixteenth Street, N. W.,
    Washington, DC 20036.
        Describes an experimental program using two-way
    television transmission via satellite. The subject of the
    program was child assessment for nurses.

Jones, P. K.; S. L. Jones; and H. L. Halliday. "Evaluation
    of Television Consultations Between a Large Neonatal
    Care Hospital and Community Hospital," Medical Care,
    January 18, 1980, p. 11-16. Philadelphia, PA. Pub-
    lished in London by Pittman Medical Company.
        Report of a $2\frac{1}{2}$-year study evaluating two-way tele-
    vision consultations between community hospital nurses
    and neonatologists at a nearby hospital. The data indi-
    cated that television consultations facilitated formation of
    the appropriate criteria for interhospital transfer. Rou-
    tine clinical screening tests seemed to be performed more
    consistently following initiation of the interhospital con-
    sultations.

Krainin, Stanley; W. Scott Andrus; and Kenneth T. Bird.
    An Evaluation of the Teleconsultation System Elements,
    1975. National Technical Information Service, Report
    Number PB-242 584/1SL, Springfield, VA 22161.
        Description of the electronic systems functioning in
    the telemedicine system linking the Veterans Administra-
    tion Hospital in Bedford, Massachusetts, and General
    Hospital in Boston. Eight subsystem groups transmit
    video, audio, data, diagnosis, control, transmission,
    support, and testing signals. Attention to hardware
    needs has provided a system in which every element is
    a response to a felt need and is actively used. Recom-
    mendations applicable to any interactive television system
    are included in the final section of the report.

Lasdon, Gail S. Manpower and Technology Alternatives to
    Increase Availability, Part I, 1977. National Technical

Information Service, Report Number HRP-0024671/OSL,
Springfield, VA   22161.
   A discussion of technology alternatives that are potentially useful for extending medical practice to remote
areas.   A two-way cable television system connecting a
rural hospital and two outpatient facilities is described.
An application of computer simulations and mathematical
programming for ambulatory care staffing analysis is
explained in detail.

Ledley, R. S., and others.   "TEXAC -- A Texture Analysis
   Computer and Its Biomedical Applications," Proceedings
   of the Fourth International Joint Conference on Pattern
   Recognition.   University of Kyoto, Kyoto, Japan.   International Association of Pattern Recognition Conference,
   November 7-10, 1978.   Bibliographic Retrieval Service,
   INSPEC Data Base Number C79030875.
   A description of the auxiliary computer system TEXAC
(Texture Analysis Computer), which is capable of per-
forming most whole-picture operations at a rapid television rate of 1/30 second.   The architecture of TEXAC
hardware and software systems, and potential biomedical
applications, are discussed.

Marquis, Jacques A.   Two-way Closed Circuit Television
   Offers Educational Programs, 1975.   National Technical
   Information Service, Report Number HRP-0007560/6SL,
   Springfield, VA   22161.
   A description of INTERACT (Interactive Television
Network), a two-way communication system among sev-
eral health care institutions in Vermont and New York.
The system began full operation in 1975, averaging 60
hours of use per week.

Maxmen, J. S.   "Telecommunications in Psychiatry," American Journal of Psychotherapy, 32(3), (July 1978):450-6.
   Association for the Advancement of Psychotherapy, 114
   East 78th Street, New York, NY 10021.
   Uses, advantages, and limitations of interactive television and videotelephones for psychiatric service and
training are discussed.

Page, G., and others.   "Teleradiology in Northern Quebec,"
   Radiology, 104(2), (August 1981):361-66.   Radiological
   Society of North America, Inc., Oak Brook Regency
   Towers, 1415 West 22nd Street, Suite 1150, Oak Brook,
   IL   60521.

A report on a two-way television network transmitting radiographic images from Northern Quebec to Montreal. Agreement on interpretation of the radiographic images as viewed via television and on-the-site reached 93 percent after three months of regular use. Overall, the radiologists were satisfied with the television transmission of radiographic images (X rays).

Park, Ben. An Introduction to Telemedicine: Interactive Television for Delivery of Health Services, 1974. Rockefeller Foundation, 1133 Avenue of the Americas, New York, NY 10021. (ERIC Document Reproduction Service Number ED 110 028)
A history of telemedicine and a description of pioneer interactive television systems used for transmitting medical information is presented.

Park, Ben, and Rashid Bashshur. "Some Implications of Telemedicine," Journal of Communication, 25(3), (Summer 1975):161-6. Annenberg School of Communication, 3620 Walnut Street, Philadelphia, PA 19140.
This article investigates the effects of telemedicine on the doctor-patient relationship.

Phillips, D. A., and W. C. Treurniet. Man-machine Interaction in the Hermes Experiments, Volume I: Experiments; Volume II: Issues; November 1978. Communications Research Centre, Department of Communication, Ottawa, Canada. (CRC Report Number 1320-1E and Number 1320-2E)
A report on a study on man-machine interaction in the University of Western Ontario Hermes Experiments. The focus of the study was on human behavior in relation to the characteristics of the telecommunications equipment and system and on issues that focus upon a particular technology and a specific aspect of behavior.

Richardson, Molly. "Interactive CAI Through Cable Television," Educational Technology, 20(11), (November 1980):57. Educational Technology Publications, 140 Sylvan Avenue, Englewood Cliffs, NJ 07632.
The United Cerebral Palsy Association of Western New York sponsors an application of computer assisted instruction called TEL-CATCH. This system permits interactive communication with closed-circuit television signals via telephone lines. Hardware includes a keyboard terminal and cathode ray tube (television screen) for individual use. TEL-CATCH was originally designed

to provide enrichment and educational experiences to
house-bound handicapped children.

Rimer, Irving. "Cable TV as a Teach-in Tool," Public
    Relations Journal, 36(9), (September 1980):24-6. Public
    Relations Society of America, 845 Third Avenue, New
    York, NY 10022.
    The American Cancer Society presented information
about breast cancer to women via television. Data from
the experiment indicated that interactive television can
significantly strengthen popular understanding of current
issues with cost effectiveness.

Rockoff, Maine L. "Telemedicine--Communications Technol-
    ogy in Health-care," Bulletin of the ASIS, 2(1), (June/
    July 1975):21-23. American Society for Information
    Sciences, 1010 Sixteenth Street, N. W., Washington, DC
    20036.
    A discussion of the use of telecommunications tech-
nology to equalize health care by distributing the knowledge of health care personnel via technology. Cost efficiency can be obtained, and geographical distances will
not form a barrier to the reception of specialized knowledge and medical expertise, according to the author.

Sanborn, Donald E.; Ann Miller; and Arthur Naitove. "Inter-
    active Television and Attitude Shift," Psychological Re-
    ports, 39(3, part 2), (1976):1162. P. O. Box 9229,
    Missoula, MT 59807.
    Results of a study using TAM attitude shift to deter-
mine the student response exposure to interactive television.

Sanborn, Donald E., and others. "Patterns of Preference:
    Nurses' View of Continuing Education," 1974. National
    Technical Information Service, Report Number HRP-
    0016850/0SL, Springfield, VA 22161.
    Report on 124 six-item questionnaires answered by
nursing personnel. The questionnaire was concerned
with preferences for scheduled courses and methods of
delivery, including two-way television.

Sanborn, Donald E., and others. "Teaching Sign Language
    by Interactive Television," American Annals of the Deaf,
    120(1), (February 1975):58-62. Conference of Educational
    Administrators Serving the Deaf, 814 Thayer Avenue,
    Silver Springs, MD 20910.

A description of the introductory course in sign language taught to five groups of non-deaf persons via
closed-circuit television.

Seibert, Dean J. , and others. Further Development and
Expansion of a Model Interactive Television System,
1974. National Technical Information Service, Report
Number PB-242 793/8SL, Springfield, VA   22161.
A report on a model interactive television system,
including operations, programming evaluation, coordinating, scheduling, and technical descriptions.

Seibert, Dean J. New Hampshire-Vermont Medical Interactive Television Network: Development and Evaluation
of a Model Interactive Television System, 1972. National Technical Information Service, Report Number
PB-220 497/2, Springfield, VA   22161.
Report of the first year of operation of the New
Hampshire-Vermont medical interactive television network. The report includes information on research design, system operation, and evaluation.

Simon, K. , and A. Moss. "A Dynamic Random Dot Stereo-
gram-based System for Strabismus and Amblyopia Screen-
ing of Infants and Young Children, " Computers in Biology
and Medicine, 11(1), (1981):133-46. Pergamon Press,
Inc. , Journals Division, Maxwell House, Fairview Park,
Elmsford, NY   10523.
Description of the vision test used to detect the presence of strabismus ("crossed eye") or amblyopia ("lazy
eye"). A color television monitor is viewed by the
patient. Through electronic connections, resulting eye
movements are recorded by computer, and computer
analysis determines analysis and diagnosis.

"Videology and the Future Practice of Medicine, " World Journal of Psychosynthesis, 9(1), (1977):15-16. World Journal Press, Box 859, East Lansing, MI   48823.
Videology (defined as blending video technology with
the science and art of medical practice) is explained;
implications for the future practice of medicine are discussed. Telefusion (television cybernetic technique) in
psychosynthesis is also discussed.

Wittson, Cecil L. , and Reba Benschoter. "Two-way Television: Helping the Medical Center Reach Out, " American Journal of Psychiatry, 129(5), (1972):624-627.

American Psychiatric Association, 1700 18th Street, N.W.,
Washington, DC   20009.

A report on the two-way closed-circuit television used
for medical treatment and education at the University of
Nebraska Medical Center in Omaha.   Much potential
benefit derives from using this technology.   Planning for
the most appropriate system will increase the effectiveness.

Wittson, Cecil L.; D. C. Affleck; and Van Johnson.   "Two-
way Television in Group Therapy," Mental Hospitals,
November, 1961.   (Retitled Hospital and Community
Psychiatry.)   Published by the American Psychiatric
Association, 1400 K Street, N. W., Washington, DC
20005.

Report on a two-way television technique developed at
a Nebraska psychiatric institute.   The technique is helpful for dyadic and small group sessions, and the technique may enable mental health personnel to extend their
services to remote areas.

APPENDIX A: PARTIAL LIST OF USERS OF I. T. F. S. EQUIPMENT

Alabama

Board of Education of Birmingham
2015 Seventh Avenue North
Birmingham, AL   35202

City Board of Education of Huntsville
Huntsville, AL   35804

California

University of Southern California
University Park
Los Angeles, CA   90007

Marysville Joint Unified School District
504 J Street, Del Monte Square
Marysville, CA   95901

University of California
San Francisco Medical Center
1438 South Tenth Street
Richmond, CA   94804

Santa Ana Unified School District
1405 French
Santa Ana, CA   92706

Florida

Pinellas County Schools
1960 East Druid Road
Clearwater, FL   33618

School Board of Volusia County
Box 1910
Daytona Beach, FL   32015

School Board of Lee County
3308 Canal Street
Fort Myers, FL   33902

Dade County Schools
Lindsey Hopkins Building
1410 N. E. Second Avenue
Miami, FL   33132

St. Petersburg Junior College
Box 13489
St. Petersburg, FL   33733

Georgia

Emory Medical University
69 Butler Street S. E.
Atlanta, GA   30303

Georgia Institute of Technology
Center for Media Based Instruction
225 North Atlanta
Atlanta, GA   30332

Illinois

Catholic TV Network of
    Chicago
1 North Woeher Drive,
    Suite 1100
Chicago, IL   60606

Illinois Institute of Tech-
    nology
3300 South Federal Street
Chicago, IL   60616

Bradley University
Department of Electrical
    Engineering and Tech-
    nology
Peoria, IL   61606

Board of Education of Ster-
    ling Township High
    School
1608 Fourth Street
Sterling, IL   61081

Indiana

Indiana Higher Educational
    Telecommunications
    System
1100 West Michigan Street
Indianapolis, IN   46202

Indiana University--Purdue
    University at Indianapolis
630 West New York Street
Indianapolis, IN   46202

Iowa

Kirkwood Community
    College
6301 Kirkwood Blvd. S.W.
Cedar Rapids, IA   52406

Kentucky

Jefferson County Board of
    Education
3332 Newburg Road
Louisville, KY   40218

Maryland

University of Maryland
International TV Building
College Park, MD   20742

Massachusetts

Boston Catholic TV Net-
    work
39 Chapel Street
Boston, MA   02160

Michigan

University of Michigan
Room 248 West Engineer-
    ing Building
Ann Arbor, MI   48108

Board of Education of the
    City of Detroit
5057 Woodward
Detroit, MI   48909

New Jersey

Mercer County Community
    College
Department of Telecommu-
    nications
1200 Old Trenton Road
Trenton, NJ   08608

New York

Board of Cooperative Edu-
cational Services
Stamford, NY   12167

Diocesan TV Center
1345 Admiral Lane
Uniondale, NY   11533

Ohio

University of Cincinnati
Cincinnati, OH   45221

ETV Association of Metro-
politan Cleveland
4300 Brookpark Road
Cleveland, OH   44134

Greater Cleveland Hospital
1021 Euclid Avenue
Cleveland, OII   44115

Oklahoma

Oklahoma State Region for
Higher Education
Televised Instruction Pro-
gram State Capital
Oklahoma City, OK   73105

Oregon

Umatilla County Interme-
diate Education District
Box 38
Pendleton, OR   97801

South Carolina

South Carolina ETV
Columbia, SC   29250

Texas

Brazosport Independent
School District
Freeport, TX   77541

TAGER
P. O. Box 688
Richardson, TX   75080

Edgewood Independent
School District
6458 West Commerce Street
San Antonio, TX   78205

Virginia

Richmond Public Schools
2907 North Boulevard
Richmond, VA   23230

Center for Excellence
(CENTEX)
P. O. Box 158
Williamsburg, VA   23185

Canada

Board of Education for the
City of London
Television Department
London 14, Ontario
Canada

Separate School Board
Recourse Center
422 Hemlock Street
Timmins, Ontario
Canada

REFERENCES

This list was compiled with the kind assistance of
David C. Parmelee
Director of Marketing
EMCEE Broadcast Products, Inc.
P. O. Box 68
White Haven, PA 18661

Additional information concerning ITFS may be found
in these:

THE EMCEE ROUNDTABLE: ITFS in America Today, 1(1),
(February 1982). EMCEE Broadcast Products, Inc.,
P. O. Box 68, White Haven, PA 18661.

Center for Excellence, Inc. The New Dimension in Medical/
Educational/Social Services: Interactive Telecommunications. CenTeX, P. O. Box 158, Williamsburg, VA
23185. (804-229-8541)

Roth, Edith Brill. "Two-Way TV Trains Teachers," Ameri-
can Education, November 1980. U. S. Department of
Education, 400 Maryland Avenue, S. W., Washington,
DC 20202. (202-245-8907)

APPENDIX B: LOCATIONS OF PILOT PROGRAMS
IN TWO-WAY TELEVISION

☐   Practically all of these programs were developed to
explore the commercial feasibility of two-way television.   In
a few cases educational uses, in the more traditional sense,
were discussed, planned, or implemented.   Most programs
were pilot programs, operational for fairly short periods of
time (one to 18 months).   In many cases commercial firms
moved the equipment to new locations in order to test different kinds of communities and different public reactions.
Some merging, blending, and consolidating of firms and
equipment has also occurred.

The locations listed below were planned or operational at the date listed.   If a traditional educational use was
included in the planning or operation, it is noted.   Otherwise,
the pilot programs were designed for the application of interactive television for commercial interests.   This commercial
development includes many options: fire alarms, burglar
systems, business conferences, survey polling, citizen par-
ticipation/observation in community meetings, purchasing via
television/computer, community/adult offerings, recreation--
the list goes on and one.   The area of potential application of
interactive television in the future is very broad.

LOCATION: British Columbia, Vancouver--1974
FIRM: British Columbia Telephone Company

The focus of the study was to determine the market
potential for conference television using video and audio two-
way transmission.   Two site locations, Vancouver and Victoria, a distance of 57 miles apart, were chosen.   At the
beginning of the study it was thought that conference television over great distances would be appealing to paying customers because transportation time and money could be utilized
in other ways.   A single high-quality camera, placed in a
fixed location to view four people sitting around a conference
table in a fairly compact area, was selected.   The single

viewing monitor was mounted flush in the wall. The decision
about equipment placement was made to ensure full view of
all participants at all times for a conference or group setting, allowing body language communication to be observed or
"sent" at all times.

Final observations for this study indicated that the
type of camera--fixed lens or movable lens, zoom lens,
monitor, broadband transmission, and switching facilities--
needs to be specifically planned for individual customers.
Connecting central city sites or transmission between city
locations seems to be more desirable.

LOCATION: Los Gatos, California -- 1971
FIRM: TeleprompTer

The purpose was to test upstream transmission with
data and three channels for video, using the cable that carries
downstream transmission. Interference for regular service
resulted. The test provided data indicating that two cables
should be used for all two-way television transmission.
(Equipment and a change in cable design has changed this
requirement since 1971.)

LOCATION: Reston, Virginia -- 1971
FIRM: MITRE Corporation

The focus of this test was to demonstrate two-way
services available through conventional twelve-channel, one-
way cable, with a return link by telephone providing audio but
no video. This was the TICCIT System (Time Shared Interactive Computer Controlled Information Television), an example of computer-assisted interaction. The subscriber could
select whatever program the computer held through standard
television. With this use of computers and the resulting
record of programs viewed, the question of privacy violation
developed. This specific pilot project ended in 1973.

TICCIT has been widely implemented in many in-house
training programs in industry and business.

LOCATION: El Segundo, California -- 1972
FIRM: Theta Cable Company; parent companies: Telepromp-
       Ter Corporation and Hughes Aircraft

The purpose of this study was to check out equipment

in individual homes, and to observe the human engineering
aspects of console convenience and cable size. Testing services planned included merchandising, ad testing, premium TV-
audience surveys and polls, security systems, and credit
card verification. Prototype equipment was to be used; expansion of the test study to 1,000 homes was postponed in
1974. However, equipment models were selected.

LOCATION: Overland Park, Kansas -- 1972
FIRM: Telecable Corporation

        This project included education from school to home
for all housebound students. Two students were selected for
the initial study, which provided favorable results. The educational phase of this study was discontinued within one
semester.
        Another phase of the project was to test the merchan-
dising of products that could be purchased via home terminals from Sears, Roebuck and Company. The project was
discontinued in 1974.

LOCATION: Carpentersville, Crystal Lake, Illinois -- 1972
FIRM: LVO Cable of North Illinois, Inc.

        The purpose of this study was to test consumer serv-
ices with specific equipment. Terminals made by Scientific
Atlanta were installed and operated by the Oak Security Company. Six active two-way terminals were installed in suburban homes. Services provided included fire, burglary, and
"panic button" alarms. Services were provided for initial
payment plus a monthly fee over and above the cable service
charge. In the year of operation of the test no fire, theft
or panic situation occurred in the houses with terminals.

LOCATION: Monroe, Georgia -- 1972
FIRM: Monroe Water, Gas and Light Company

        This system, municipally owned through the local
utility company, was completed in September 1972. Two-
way operation of 21 channels downstream using a low-band
return of 5-35 MHz was installed. Security alarm systems
were to be installed. Involvement of public schools, through
originating programming from the school buildings, was
planned.

LOCATION: Mesa, Arizona -- 1973
FIRM: TOCOM

Through an arrangement with Rossmoor's Leisure
World, between 40 and 200 homes were to be wired for interactive services, including burglar detection, fire, and medical emergency alarms. Pay TV, opinion polling, merchandising, meter reading, remote video origination, and remote
control of home devices were also planned. In 1983, between
30 and 40 television channels were available for subscribers,
as well as an emergency "button" to electronically signal for
assistance with health, security, or fire problems.

LOCATION: Irving, Texas -- 1973
FIRM: Total Communications of Irving; parent company:
        Leacom, Inc.

The purpose of this study was to test urban-dweller
preferences in cable television, determine the amount of prof-
it that could be expected, and examine the operating problems connected with two-way services.
    This location receives good one-way television recep-
tion without cable.   There were 1,500 terminals used in the
initial study.   Cable services were offered without charge,
including two-way and premium channels.   Merchandising,
meter reading, and opinion polling were offered; firms desir-
ing these services were charged for cable use.

LOCATION: Orlando, Florida -- prior to 1973
FIRM: Orange Cablevision, Inc.; parent company: American
      Television Communications (ATC)

This study was designed primarily to test equipment.
Electronic Industrial Engineering (EIE) built equipment that
"performed perfectly" for burglary, fire and emergency
alarms, meter reading, pay TV, merchandising, polling, and
credit card verification.
    Another study--in 1974--conducted in the same
location by the same firm examined a test market for services such as pay TV, merchandising, ad testing, security
systems, and data transmission; the potential for educational
services for an additional fee was also explored.   Point-to-
point services were to interconnect the major teaching hospitals with satellite hospitals and a nursing home.   Placement
of terminals was to be by subscription, with the type of
terminal dependent upon the services desired.

LOCATION: Stockton, California -- 1975
FIRM: MITRE Corporation

This test situation developed from the data MITRE
Corporation gained from Reston, Virginia; California; and 31
other community evaluations. Stockton provided the most
useful cable system and the demographics MITRE wanted in
a test location. The cable system was a fully active bidirec-
tional dual cable system, one of very few in the United States
at that time. Prototype services were to be developed in
many categories: 1) instruction--elementary grade, computer-assisted; 2) companionship for the housebound (communication services for the elderly, idle, isolated, etc.);
3) community ombudsman--providing storage retrieval for
information about the community and all levels of government;
4) programming originating in schools, and transmitted into
the homes for community information; 5) entertainment--
including interactive computer games; 6) community programming, for discussion, public forum, and citizen awareness;
and 7) health care. The services initially were to be free,
then provided on a subscription basis. However, delays
occurred and funding was indefinite. Many tests were conducted on a limited basis, providing input about equipment
and allowing model revisions and improvements.

LOCATION: Chaska, Minnesota -- 1975

The Public Schools of Chaska were the site for a
proposed three-year study to evaluate the effectiveness of
two-way interactive cable television in a secondary school
setting. Four subject areas were taught to 25 to 30 high
school students via two-way television. Three experimental
groups and one control group were used. First-year reports
seemed favorable, but funding for the remaining two years of
the study was not granted.

LOCATION: Remote sites in the Appalachian Mountains
TITLE: Appalachian Mountain Project -- 1975

This was a modified two-way television study in that
interaction was by delayed audio response, using telephone
connections after the completion of the televised program.
Two schools in remote areas were connected in the initial
study, which provided data that led to the development of the
Appalachian Community Satellite Network (ACSN). ACSN

provides hundreds of hours of educational programming in
"live" transmission format to many young adults throughout
the southeastern United States. Through videotape many of
these courses are viewed in other parts of the country.

LOCATION: Sites within the states of Colorado, Idaho, Mon-
           tana, New Mexico, Utah, and Wyoming; project
           headquarters: Denver, Colorado. Locations
           within 18 additional states were used for spe-
           cific test situations.
TITLE: Application Technology Satellite-6 (ATS-6)

     This project explored communication technology util-
izing the NASA ATS-6. The Application Technology Satellite-
6 was first launched May 30, 1974. Six independent experiments grew out of this project. These were the Indian Health
Service (IHS); Washington, Alaska, Montana, Idaho Program
(WAMI); Alaskan Educational Experiment; Appalachian Educational Satellite Project (AESP); and the Satellite Technology
Demonstration (STD). Major project objectives were to test
the feasibility of a satellite-based media distribution system
for isolated, rural populations and to test and evaluate user
acceptance and the cost of various delivery modes using a
variety of materials.

LOCATION: Saskatchewan and Quebec Provinces, Canada --
           1977
TITLE: SASKEBEC

     This was a three-month project designed to link two
provinces in Canada through two small communities, Zenon-
Park, Saskatchewan, and Baie St. Paul, Quebec. The project was designed so that the community needed to organize
and help the television project be successful. Two purposes
were involved: 1) to help a small, isolated French-speaking
community preserve its language and culture and 2) to observe whether attitudes in the communities were modified by
this television project.

LOCATION: Columbus, Ohio -- began in December 1977.
TITLE: QUBE, a division of Warner Amex Corporation, Inc.

     QUBE is a cable subscription channel, providing pro-
gramming on 30 or more channels. Many programs involve

audience response through audience survey, polling, and
merchandising.   The QUBE concept is steadily expanding to
other cities in Ohio and throughout the United States.

LOCATION: Local government offices in the New York, New
            Jersey, and Connecticut metropolitan region --
            1969 to 1974
TITLE: MRC-TV System   (Metropolitan Regional Council
            Television System)
FUNDING: National Science Foundation

      This was the first intergovernmental teleconferencing
system, which linked the chief elected officials in the New
York, New Jersey, and Connecticut metropolitan region.   The
system included two-way microwave television transmission
that connected the MRC regional headquarters with nine primary stations located in county seats and with other selected
cities in the teleconferencing system.

LOCATION: Albany Medical College, Albany, New York,
            connecting with over 20 medical stations in
            eastern New York and western Massachusetts
TITLE: Albany Medical College interconnect system

      This system began in 1955 as an amateur band radio
network designed to connect medical personnel via audio; its
purpose was to provide consultation services to remote medical facilities.   Through the years, the amateur radio network, staffed by volunteers, became more sophisticated, and
two-way television transmission developed.   The two-way
video and audio connections have been utilized for a number
of years.   Medical education for practitioners and   nursing
staff has been added to the interactive programming.

LOCATION: Hospitals and universities in the New England
            area -- 1972
TITLE: New England Microwave Network

      This full two-way interactive television system used
van-mounted microwave units to connect several hospitals
and clinics in New Hampshire and Vermont.   This experimental microwave television network placed primary emphasis on site-to-site service to small hospitals from university
facilities.   Dartmouth Medical School, The University of

Hanover, Claremont General Hospital, the University of Vermont, and Central Vermont Hospital were included in the network.

LOCATION: Headingley, Manitoba, Canada (a suburb of
                       Winnipeg) -- 1980
TITLE: Project IDA
FUNDING: Manitoba Telephone System (MTS)

This project is named IDA in memory of Manitoba's first female telephone operator, who provided prompt connections with other telephone users as well as news, information, counsel, and emergency service. Through electronic wiring, MTS plans to provide these same services to customers during the 1980's. The project, designed initially for 100 clients, will link customers with many computer-response two-way services. Fire alarms, pay TV, meter reading, videotext services, energy management, and opinion polling are features included for the initial phase of Project IDA. As the program develops, data will determine the feasibility of continuing or expanding programs.

REFERENCE ADDRESSES

British Columbia Telephone
    Company
37777 Kingsway
Burnaby, B.C. V5H 3Z7
Canada

OR

826 Yates Street
Victoria, B.C. V8W 2H9
Canada

Teleprompter Cable System,
    Inc.
8561 Nuevo Avenue
Fontana, CA 92335

MITRE Corporation
1820 Dolly Madison Boulevard
McLean, VA 22102

Telecable Corporation
    (Headquarters)
P. O. Box 720
740 Duke Street
Norfolk, VA 23510

LVO Cable of North Illinois, Inc.
LVO Cable, Inc. United
    Cable TV Corporation
300 Carpenter Boulevard
Carpentersville, IL 60110

Monroe Water, Gas and
    Light Company
215 North Broad
Monroe, GA 30655

TOCOM Inc.
3301 Royalty Row
Irving, TX 75062

Leacom Inc.
Contact: Leacom Cable
   Vision
c/o Communications Systems,
   Inc.
P. O. Box 47-417
Truth or Consequences,
   NM 87901

MRC-TV System
Metropolitan Regional Coun-
   cil, Inc.
Suite 2437, One World
   Trade Center
New York, NY 10048

Albany Medical College
   Interconnect
Albany Medical College
New Scotland Avenue
Albany, NY 12208
(518-445-3125)

New England Microwave
   Network
Dartmouth College
Hanover, NH 03755

Project IDA
Manitoba Telephone System
Box 6666, 489 Empress
   Street
Winnipeg, Man. R3C 3V6
Canada

AV Director
Independent School District
   #112
Chaska, MN 55318

Appalachian Mountain Proj-
   ect
Dr. Norfett Williams
University of Kentucky
Lexington, KY 40506

Satellite Technology Dem-
   onstration
Federation of Rocky Moun-
   tain States, Inc.
Denver, CO 80201
(Sponsor: National Institute
   of Education,
DHEW,
1200 19th Street, N. W.
Washington, DC 20208)

SASKEBEC
University of Regina,
Regina, Sask. S4S-OA2
Canada

QUBE
930 Kinnear Road
Columbus, OH 43212

American Television and
   Communications, Inc.
160 Inverness West
Englewood, CO 80112
(303-799-1200)

Information concerning additional pilot programs using
interactive television may be found in the following:

Leduc, N. F. ; C. D. Shepard; E. McFarlane; and J. Costa.
   The Development of Telecommunications Services: A
   Review of Projects, Volume 11, October 1979. Trip
   Reports, Department of Communications, Canada. Con-
   tact the Department of Communications, 300 Slater Street,
   Ottawa, Ont. K1A OC8, Canada. (613-995-8185)

APPENDIX C: DIRECTORS AND COORDINATORS

☐   Information in this text was enhanced and, in some cases, clarified, by the assistance of many concerned professionals. Sincere appreciation is extended to the following individuals for sharing information and ideas related to interactive television communication.

Fran Davis, Superintendent
Morning Sun Public Schools
Morning Sun, IA   52640

Dale Hemmie, President
Telecom Engineering Inc.
Fort Madison, IA   52627

Richard Lundgren, Principal
Eagle Bend School District
  #790
Eagle Bend, MN   56446

Will James, Superintendent
Eagle Bend Independent School
  District #790
Eagle Bend, MN   56446

Rodney Moen, General Manager
Western Wisconsin Communications Cooperative
202 Whitehall Road, P.O.
  Box 326
Independence, WI   54747

Ellworth Beckman, Jr.,
  Project Director
Project CIRCUIT
13th and Francis Street
Osseo-Fairchild High School
Osseo, WI   54758

Samuel M. Genensky, Ph.D.,
  Director
Center for the Partially
  Sighted
Santa Monica Hospital Medical Center
1250 Sixteenth Street
Santa Monica, CA   90404

Tora Bikson, Ph.D.
The Rand Corporation
1700 Main Street
Santa Monica, CA   90406

Levonne Kelley
19100 East Killian Street
Rowland Unified School District
Rowland Heights, CA   91745

James von Feldt, Ph.D.,
  President
VONTECH, INC.
1320 Buffalo Road
Rochester, NY   14624

Dr. Thomas Baldwin
Department of Telecommunication
Michigan State University
East Lansing, MI   48824

Dr. Judith Bazemore
901 Lucerne Drive
Spartanburg, SC   29302

Dr. Robert Fina
Kutztown State College
Kutztown, PA   19530

Earl Haydt, Regional Manager
ATC (American Telecom-
munications Corporation)
R. D. 5, Route 422 West
Sinking Springs, PA   29608

Dr. Richard West, Assistant
Vice-president
University of California
Systemwide Administration
Berkeley, CA   94720

Dr. Ralph D. Mills, Dean
Extended Education
The California State Uni-
versity
Office of the Chancellor
Long Beach, CA   90802

Dr. John P. Witherspoon,
Project Director
The Public Telecommuni-
cations Project: Plan-
ning for the Decade
Center for Communications
San Diego State University
San Diego, CA   92182

J. Patrick Loughboro,
President
JP Associates, Inc.
3115 Kashiwa Street
Torrance, CA   90505

Dr. Charles Urbanowicz
Associate Dean
Regional and Continuing
Education
California State University,

Chico
Chico, CA   95929-0250

Dr. Ralph Weintraub, Direc-
tor
Instructional Media Center
California State University,
Chico
Chico, CA   95929-0250

Dr. Judith Lemon
Stanford Instructional Tele-
vision Network
Room 343 Durand Building
Stanford University
Stanford, CA   94305

Peter C. Kerner, Coordinator
Educational Television Office
University of California,
Berkeley
Berkeley, CA   94720

Craig Ritter, Director
IUSD 2-Way Television
Project
Irvine Unified School District
2941 Alton Avenue/P. O. Box
19535
Irvine, CA   92713

Arthur D. Shmarak, Manager
Health Sciences Television
University of California,
Davis
Davis, CA   95616

Dr. William Shipp
Brown University
P. O. Box 1857
Providence, RI   02912

David C. Parmelee
Director of Marketing
EMCEE Broadcasting Prod-
ucts, Inc.
P. O. Box 68
White Haven, PA   18661

Dr. Mel Mickerson
California State College,
  Stanislaus
P. O. Box 1000
Turlock, CA 95380

Dr. Morton D. Miller
Director, Instructional
  Media
Room 19 Olson Hall
University of California,
  Davis
Davis, CA 95616

Dr. Jack Munushian
Professor of Electrical En-
  gineering and Director of
  Instructional Television
  Network
School of Engineering
University of Southern Cali-
  fornia, Los Angeles
Los Angeles, CA 90089

John A. Curtis
Box 158
Center for Excellence, Inc.
Williamsburg, VA 23187

Sister T. Kevin
St. Mary's School for the
  Visually Impaired
Merrion Road
Dublin 4
Ireland

Lisa Navarro
Educational Television
  Center
CTN/San Francisco
324 Middlefield Road
Menlo Park, CA 94025

Marc Spenzel
Visual Tek
1610 26th Street
Santa Monica, CA 90404

art 8
art appreciation   21
art history   21
Bachelor of Arts in Social Sciences   65
business management   55, 65
Chinese studies   61
computer science   19, 21, 55, 57, 65
creative writing   4
education   65
employment skills   8
engineering   55, 57, 65
English literature   61
exotic languages   61
faculty in-service programs   71
French   16, 21, 48
general interest subjects   55
German   8, 21
history   21
instrumental music (brass, woodwind)   21
Latin American studies   61
management development   55
mass communications   8
Master of Arts in School of Engineering   61
MBA   55
medical (autopsy rooms, operating, patient examining)   60
music theory   19, 21
oceanography   47, 48
paralegal   65
physics   55
physiology   4
political science   61
psychology   65
Russian history   61
Russian language   48
shorthand   21, 22
Slavic languages   61
Spanish   4, 19, 21
statistics   55
team teaching   71
Course offerings (non-credit)
adult education   5, 9, 36
aging   16
business   39
Chamber of Commerce   47
childrearing   36
community interest programs   17, 19
computer equipment   36